Gearing
A Mechanical Designers' Workbook

Other Books in
The McGraw-Hill Mechanical Designers' Workbook Series

MACHINE DESIGN FUNDAMENTALS: A Mechanical Designers' Workbook

DISTORTION AND STRESS: A Mechanical Designers' Workbook

CORROSION AND WEAR: A Mechanical Designers' Workbook

FASTENING, JOINING, AND CONNECTING: A Mechanical Designers' Workbook

MECHANISMS: A Mechanical Designers' Workbook

BEARINGS AND LUBRICATION: A Mechanical Designers' Workbook

POWER TRANSMISSION ELEMENTS: A Mechanical Designers' Workbook

Gearing

A Mechanical Designers' Workbook

Editors in Chief

Joseph E. Shigley

Professor Emeritus
The University of Michigan
Ann Arbor, Michigan

Charles R. Mischke

Professor of Mechanical Engineering
Iowa State University
Ames, Iowa

McGraw-Hill Publishing Company

New York   St. Louis   San Francisco   Auckland   Bogotá
Caracas   Hamburg   Lisbon   London   Madrid   Mexico
Milan   Montreal   New Delhi   Oklahoma City
Paris   San Juan   São Paulo   Singapore
Sydney   Tokyo   Toronto

Library of Congress Cataloging-in-Publication Data

Main entry under title:

Standard handbook of machine design.

    Includes index.
    1. Machinery—Design—Handbooks, manuals, etc.
I. Shigley, Joseph Edward. II. Mischke, Charles R.
TJ230.S8235 1986   621.8′15   85-17079
ISBN 0-07-056892-8
ISBN 0-07-056926-6 (workbook)

Copyright © 1990 by McGraw-Hill, Inc. All rights reserved. Printed in the United States of America. Except as permitted under the United States Copyright Act of 1976, no part of this publication may be reproduced or distributed in any form or by any means, or stored in a data base or retrieval system, without the prior written permission of the publisher.

1234567890 KGP/KGP 8965432109

ISBN 0-07-056926-6

The material in this volume has been published previously in *Standard Handbook of Machine Design* by Joseph E. Shigley and Charles R. Mischke. Copyright © 1986 by McGraw-Hill, Inc. All rights reserved.

The editors for this book were Robert Hauserman and Scott Amerman, and the production supervisor was Dianne Walber. It was set in Times Roman by Techna Type.

Printed and bound by The Kingsport Press.

Information contained in this work has been obtained by McGraw-Hill, Inc., from sources believed to be reliable. However, neither McGraw-Hill nor its authors guarantees the accuracy or completeness of any information published herein and neither McGraw-Hill nor its authors shall be responsible for any errors, omissions, or damages arising out of use of this information. This work is published with the understanding that McGraw-Hill and its authors are supplying information but are not attempting to render engineering or other professional services. If such services are required, the assistance of an appropriate professional should be sought.

For more information about other McGraw-Hill materials, call 1-800-2-MCGRAW in the United States. In other countries, call your nearest McGraw-Hill office.

In Loving Memory of
Opal Shigley

CONTENTS

Contributors   ix
Preface   xi

1. Power Screws, *by William C. Orthwein*   **1**

    1-1   Thread Profiles   3
    1-2   Initial Design Considerations   3
    1-3   Buckling   5
    1-4   Thread Stresses   6
    1-5   Lead-Angle Selection   10
    1-6   Efficiency, Lubrication, and Backlash   10
    1-7   Ball Screws   14
    1-8   Reversing Screws   19
    1-9   Roller Screws   19
           REFERENCES   22

2. Spur Gears, *by Joseph E. Shigley*   **23**

    2-1   Definitions   24
    2-2   Tooth Dimensions and Standards   27
    2-3   Force Analysis   27
    2-4   Fundamental AGMA Rating Formulas   28
           REFERENCES   34

3. Bevel and Hypoid Gears, *by Theodore J. Krenzer and Robert G. Hotchkiss*   **35**

    3-1   Introduction   36
    3-2   Terminology   36
    3-3   Gear Manufacturing   41
    3-4   Gear Design Considerations   44
    3-5   Gear-Tooth Dimensions   55
    3-6   Gear Strength   69
    3-7   Design of Mountings   77
    3-8   Computer-Aided Design   94

4. Helical Gears, *by Raymond J. Drago* — **107**

- 4-1 Introduction — 108
- 4-2 Types — 109
- 4-3 Advantages — 109
- 4-4 Geometry — 111
- 4-5 Load Rating — 116
- REFERENCES — 178

5. Worm Gearing, *by K. S. Edwards* — **179**

- 5-1 Introduction — 181
- 5-2 Kinematics — 182
- 5-3 Velocity and Friction — 184
- 5-4 Force Analysis — 184
- 5-5 Strength and Power Rating — 193
- 5-6 Heat Dissipation — 195
- 5-7 Design Standards — 196
- 5-8 Double-Enveloping Gear Sets — 205
- REFERENCES — 210

Index 211

CONTRIBUTORS

Raymond J. Drago, *Senior Engineer,* Advanced Power Train Technology, Boeing Vertol Company, Philadelphia, Pa.

K. S. Edwards, *Professor of Mechanical Engineering,* The University of Texas at El Paso, Tex.

Robert G. Hotchkiss, *Director,* Gear Technology, Gleason Machine Division, Rochester, N.Y.

Theodore J. Krenzer, *Manager,* Gear Theory Department, Gleason Machine Division, Rochester, N.Y.

William C. Orthwein, *Professor,* Southern Illinois University at Carbondale, Carbondale, Ill.

Joseph E. Shigley, *Professor Emeritus,* The University of Michigan, Ann Arbor, Mich.

PREFACE

There is no shortage of good textbooks treating the subject of machine design and related topics of study. But the beginning designer quickly learns that there is a great deal more to successful design than is presented in textbooks or taught in technical schools or colleges. A handbook connects formal education and the practice of design engineering by including the general knowledge required by every machine designer.

Much of the practicing designer's daily informational needs are satisfied in various pamphlets or brochures such as are published by the various standards organizations as well as manufacturers of various components used in design. Other sources include research papers, design magazines, and corporate publications concerned with specific products. More often than not, however, a visit to a design library or to a file cabinet will reveal that a specific publication is on loan, lost, or out of date. A handbook is intended to serve such needs quickly and immediately by giving the designer authoritative, up-to-date, understandable, and informative answers to the hundreds of such questions that arise every day in the work of a designer.

The *Standard Handbook of Machine Design*[*] was written for working designers, and its place is on their desks, not on their bookshelves, for it contains a great many formulas, tables, charts, and graphs, many in condensed form. These are intended to give quick answers to the many questions that seem to arise constantly.

The *Mechanical Designers' Workbook* series consists of eight volumes, each containing a group of related topics selected from the *Standard Handbook of Machine Design*. Limiting each workbook to a single subject area of machine design makes it possible to create a thin, convenient volume bound in such a manner as to open flat and provide an opportunity to enter notes, references, graphs, equations, standard corporate practices, and other useful data. In fact, each chapter in every workbook contains gridded pages located in critical sections for this specific purpose. This flat-opening workbook is easier to use on the designer's work space and will save much wear and tear on the source handbook.

Chapter 1 of *Gearing* deals with power screws. Spur gears is the subject of Chapter 2, but it also serves as an introduction to gearing in general by explaining the basic concepts, definitions, nomenclature, and the charts that are used in designing many types of gears.

Chapters 3 and 4 contain much important new material and reveal fully the design methods used by those working at the edge of knowledge in these fields. Chapter 3 deals with bevel and hypoid gearing, and Chapter 4 with helical gearing. The subject of worm gearing is that of Chapter 5, which concludes this volume.

Care has been exercised to avoid error. The editors will appreciate being informed of errors discovered so that they may be eliminated in future printings.

JOSEPH E. SHIGLEY
CHARLES R. MISCHKE

[*]By Joseph E. Shigley and Charles R. Mischke, Coeditors-in-Chief, McGraw-Hill Publishing Company, New York, 1986.

ABOUT THE EDITORS

Joseph E. Shigley has had a long and distinguished career both as an educator and as a consultant in machine design and mechanical engineering. He was a professor of mechanical engineering at the University of Michigan for 21 years. He is a Fellow of the American Society of Mechanical Engineers. He received the ASME Mechanisms Committee Award in 1974, the Worchester Reed Warner Medal in 1977, and the Machine Design Award in 1985. He is the author or co-author of eight McGraw-Hill books.

Charles R. Mischke is Professor of Mechanical Engineering at Iowa State University, a consultant to industry, contributor of many technical papers, author or co-author of five books, and co-editor of the *Standard Handbook of Machine Design*. He received the Ralph Teeter Award of the Society of Automotive Engineers in 1977, and the University's Outstanding Teacher Award in 1980. A Fellow of the American Society of Mechanical Engineers (ASME), he also serves on its Reliability, Stress Analysis, and Failure Prevention Executive Committee.

chapter 1
POWER SCREWS

WILLIAM C. ORTHWEIN, Ph.D.
Professor
Southern Illinois University at Carbondale
Carbondale, Illinois

GLOSSARY OF SYMBOLS

A Area
b Flat width
C_i Percentage of ball-screw life at load F_i
d_i Ball diameter (ball screw)
d_n Minor diameter of the nut
d_p Pitch diameter
d_s Minor diameter of the screw
D Major diameter
D_N Nut outside diameter (across the flats or at the root, circle of a gear nut)
D_n Major diameter of internal thread (nut)
E Elastic modulus
F Force
h Thread height or depth
l Unsupported length
L Lead
L_e Length of external thread engagement
n Number of threads per unit of length ($n = 1/p$)
N Number of turns per unit of length ($N = 1/L$)
N_e Number of threads engaged
p Pitch
r_c Collar mean radius (average radius)
r_p Screw pitch radius ($r_p = d_p/2$)
r_s Minor radius of the screw
S_y Yield strength
T Torque

w	Width of external thread at root
X	Ball-screw life in units of length of travel
Z	Number of load-carrying balls per turn
α	Half angle of a symmetrical thread
α_1	Angle of leading thread flank
β	Circumferential-stress geometry factor
δ	Backlash
ζ	End-condition factor in buckling
η	Efficiency
λ	Lead angle
μ_c	Collar coefficient of friction
μ_t	Thread coefficient of friction
ξ	Bearing-stress geometry factor
ρ	Thread-friction parameter
σ	Normal stress
σ_b	Bearing stress
τ	Shear stress
ϕ	Projection angle

1-1 THREAD PROFILES

Basic thread forms for power screws are shown in Fig. 1-1. No standard exists for the *square thread* shown in Fig. 1-1a, nor a modification (not shown); this thread is useful in some applications, however, and is easy to analyze.

The *Acme thread*, shown in Fig. 1-1b, has a thread angle $\alpha = 14\frac{1}{2}$ degrees. This standard thread is classified as *general purpose* or *centralizing*, depending on the tolerances (see Refs. [1-2] and [1-3]). Use Table 1-1 to select the pitch corresponding to any size. Formulas for the basic width of the flat are listed in Table 1-2.

The *buttress thread* of Fig. 1-1c is also standardized, and the sizes and most of the pitches shown in Table 1-1 apply. Complete details of this profile may be found in ANSI standard B1.8-1977 [1-3]. The threads in Fig. 1-1 are all single-start and right-hand. Refer to [1-1] for screw-thread terminology.

1-2 INITIAL DESIGN CONSIDERATIONS

Selection of the minimum root diameter of a power screw is the first step in power-screw design. It must be large enough to prevent buckling if the screw is axially loaded and large enough to prevent excessive stress and deflection if the screw is laterally loaded. Analysis of the critical speed is seldom necessary for power-screw design.

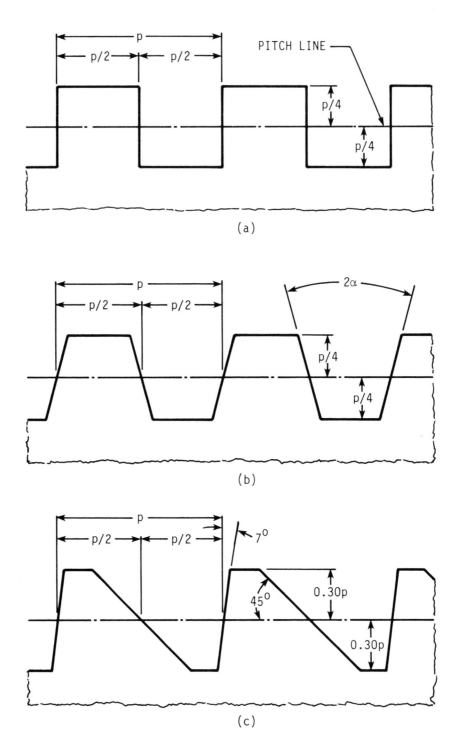

FIG. 1-1 Basic thread forms. (*a*) Square; (*b*) Acme; (*c*) buttress. The Acme thread is also available as a *stub Acme* in which the thread height is 0.30*p*. Two modified stub Acmes are available with thread heights of 0.375*p* and 0.250*p*. To allow for clearances and fillets or rounding, the actual thread heights for external and internal threads will deviate somewhat from the theoretical values shown.

TABLE 1-1 Standard Thread Sizes for Acme Threads†

Size D, in	Threads per inch n
$\frac{1}{4}$	**16**
$\frac{5}{16}$	16, **14**
$\frac{3}{8}$	16, 14, **12, 10**
$\frac{7}{16}$	16, 14, **12, 10**
$\frac{1}{2}$	16, 14, 12, **10**, 8
$\frac{5}{8}$	16, 14, 12, 10, **8**
$\frac{3}{4}$	16, 14, 12, 10, 8, **6**
$\frac{7}{8}$	14, 12, 10, 8, **6**, 5
1	14, 12, 10, 8, 6, **5**
$1\frac{1}{8}$	12, 10, 8, 6, **5**, 4
$1\frac{1}{4}$	12, 10, 8, 6, **5**, 4
$1\frac{3}{8}$	10, 8, 6, 5, **4**
$1\frac{1}{2}$	10, 8, 6, 5, **4**, 3
$1\frac{3}{4}$	10, 8, 6, **4**, **4**, 3, $2\frac{1}{2}$
2	8, 6, 5, **4**, 3, $2\frac{1}{2}$, 2
$2\frac{1}{4}$	6, 5, **4**, 3, $2\frac{1}{2}$, 2
$2\frac{1}{2}$	5, **4**, 3, $2\frac{1}{2}$, 2
$2\frac{3}{4}$	**4**, 3, $2\frac{1}{2}$, 2
3	**4**, 3, $2\frac{1}{2}$, 2, $1\frac{1}{2}$, $1\frac{1}{3}$
$3\frac{1}{2}$	**4**, 3, $2\frac{1}{2}$, 2, $1\frac{1}{2}$, $1\frac{1}{3}$, 1
4	**4**, 3, $2\frac{1}{2}$, 2, $1\frac{1}{2}$, $1\frac{1}{3}$, 1
$4\frac{1}{2}$	3, $2\frac{1}{2}$, 2, $1\frac{1}{2}$, $1\frac{1}{3}$, 1
5	3, $2\frac{1}{2}$, 2, $1\frac{1}{2}$, $1\frac{1}{3}$, 1

†The preferred size is shown in boldface.

1-3 BUCKLING

The subject of buckling is treated extensively elsewhere. Here we present only the gist of the topic. Adapting the Euler equation to power screws gives

$$\frac{F}{A} = \frac{\pi^2 E}{4(\zeta l/r_s)^2} \tag{1-1}$$

where r_s = root radius. This equation contains no factor of safety, and hence F is the *buckling* or *critical load*. Of course, the Euler equation applies only when the l/r_s ratio is large, that is, when

$$\frac{l}{r_s} > \frac{\pi}{\zeta}\left(\frac{E}{2S_y}\right)^{1/2} \tag{1-2}$$

For smaller l/r_s ratios, the J. B. Johnson or parabolic formula can be used. For power

TABLE 1-2 Basic Width of Flat for Acme Threads

Thread form	Thread height h	Flat width b
Acme	$0.50p$	$0.3707p$
Stub Acme	$0.30p$	$0.4224p$
Stub Acme	$0.25p$	$0.4030p$
Stub Acme	$0.375p$	$0.4353p$

screws, this formula can be written

$$\frac{F}{A} = S_y \left[1 - \left(\frac{\zeta l}{\pi r_s}\right)^2 \frac{S_y}{E} \right] \tag{1-3}$$

where again F is the buckling or critical load as defined by this equation. Values of the end-condition constant ζ in these three equations are given in Fig. 1-2.

1-4 THREAD STRESSES

The *bearing stress* on the screw thread or nut thread can be obtained from the equation

$$\sigma_b = -\frac{4F\xi}{\pi N_e(D^2 - d_n^2)} \tag{1-4}$$

where the minus sign indicates compression. The factor ξ is used to account for the effects of the flank angle α_1, the lead angle λ, and the coefficient of friction μ_t between the threads. This factor is obtained from the equation

$$\xi = \frac{\cos \alpha_1}{(1 + \tan^2 \alpha_1 \cos^2 \lambda)^{-1/2} - \mu_t \tan \lambda} \tag{1-5}$$

It is worth noting that the error of using $\xi = 1$ when $\mu_t < 0.10$ and $\lambda < 5$ degrees is less than 1.0 percent. Values of the coefficients of friction may be obtained from Table 1-3.

The active length of the nut is the overall length minus the chamfer allowance and is

$$L_e = N_e p \tag{1-6}$$

The *bending stress* at the thread root may be approximated by representing the thread as a cantilever of width $\pi N_e d_p$ and depth w and supporting a concentrated load F at a distance $(d_p - d_s)/2$ from the fixed end. The flexural formula then gives the bending stress as

$$\sigma = \frac{3F(d_p - d_s)}{\pi d_p N_e w^2} \tag{1-7}$$

The distance w can be obtained from Table 1-2 as $w = p - b$ for Acme threads. Figure 1-3 can be used to learn that $w = 0.9072p$ for flat-root buttress threads. This is also a close approximation for round-root buttress threads.

Notes · Drawings · Ideas

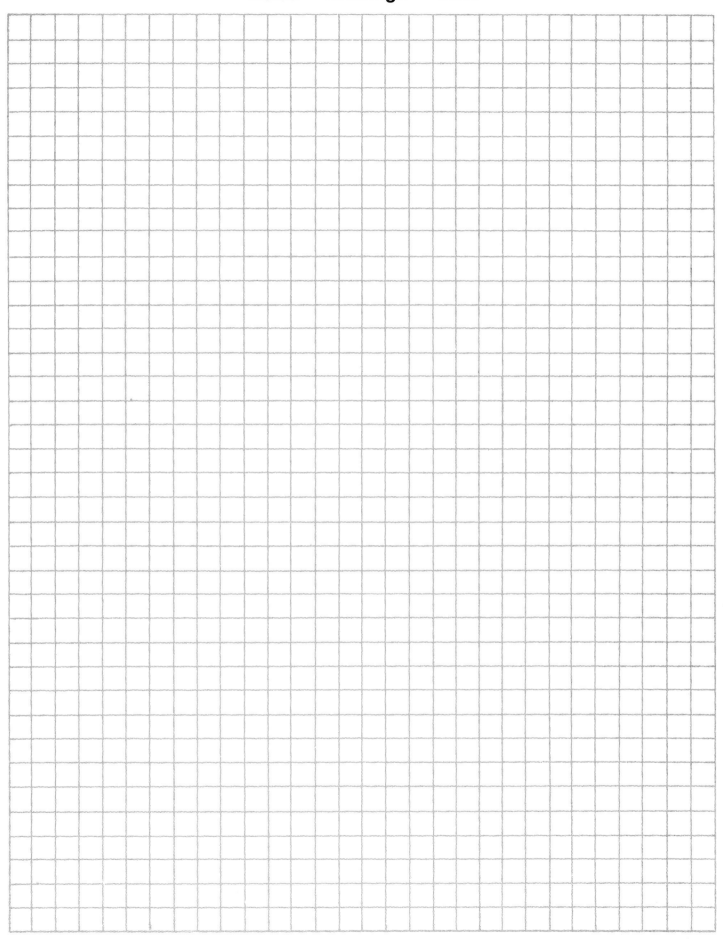

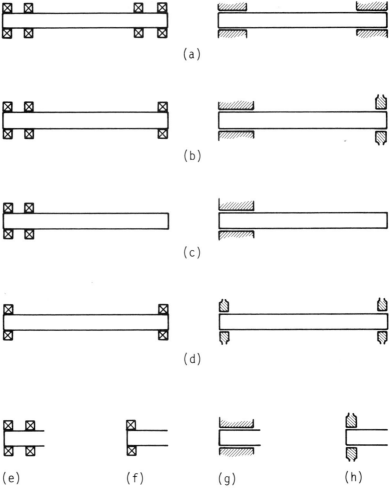

FIG. 1-2 Column end conditions for power screws. (*a*) Both ends fixed, $\zeta = 0.50$; (*b*) one end fixed and one end supported, $\zeta = 0.707$; (*c*) one end fixed and one end free, $\zeta = 2$; (*d*) both ends supported, $\zeta = 1$; (*e*) a fixed end has two tapered roller bearings back-to-back, or two angular-contact ball bearings back-to-back, or two radial roller bearings with an additional thrust bearing (not shown); (*f*) a supported end is obtained with a single-row ball bearing or a spherical roller bearing; (*g*) a long journal bearing with small clearance and a rigid mount may be considered as a fixed end; (*h*) a short journal bearing with a large clearance or a flexible mount should be treated as a supported end.

TABLE 1-3 Range of Coefficient of Friction μ for Power Screws†

Contact surface	Friction coefficient
Metal to metal	0.40 to 0.80
Lubricated	0.005 to 0.20
Rolling element	0.0015 to 0.008

†These values apply to both the thread friction and the collar friction.

SOURCE: From Machine Design, Penton/IPC, Cleveland, Ohio, by permission.

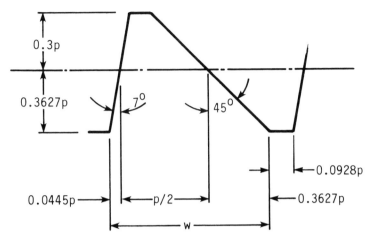

FIG. 1-3 Drawing of actual buttress external thread showing how to obtain the thread depth w for an external thread. Note that the distance from the pitch cylinder to the root cylinder differs from the theoretical value shown in Fig. 1-1c.

The *shear stress* at the root may be approximated by using an area $\pi d_s w N_c$ and the commonly accepted $\frac{4}{3}$ factor associated with a parabolic shear stress distribution across the tooth width. This gives

$$\tau = \frac{4F}{3\pi d_s w N_c} \tag{1-8}$$

The radial component of the force perpendicular to the leading flank of the screw thread induces a *circumferential* or *hoop stress* in the nut. This is a tensile stress, and it can be approximated by an equation based on pressurized cylinder theory. The equation is

$$\sigma = \frac{F\beta}{\pi d_n p N_c}\left(\frac{D_N^2 + D_n^2}{D_N^2 - D_n^2}\right) \tag{1-9}$$

where β is a coefficient used to account for the effects of the leading flank angle α_1, the lead angle λ, and the thread friction μ_t. Thus

$$\beta = \frac{\tan \alpha_1}{1 - \mu_t (\tan \lambda)(1 + \tan^2 \alpha_1 \cos^2 \lambda)^{1/2}} \tag{1-10}$$

Typical values for an SI or Unified thread form might be $\lambda = 2$ degrees, $\alpha_1 = 30$ degrees, and $\mu_t = 0.20$. This gives $\beta = 0.5820$. A corresponding buttress thread would have $\alpha_1 = 7$ degrees, giving $\beta = 0.1237$. This is additional motivation for using Acme or buttress threads for power screws.

The remarkable reduction in hoop stress, almost 80 percent, obtained by reducing α_1 from 30 to 7 degrees clearly shows the advantage of the buttress thread over the Unified thread form for power-screw applications. The disadvantage is that it is a one-way thread form in that it can exert a large force in only one direction if the hoop stress is to remain small. Although the square thread form shown in Fig. 1-1a produces no lateral force on the nut and is bidirectional, it is expensive to produce because of the machining problems in cutting the perpendicular surfaces. Acme and stub Acme thread forms are a compromise to achieve equal performance in both directions with reasonable production costs. Even the 5-degree flank angle used in the modified square thread opens up the root enough to a cutting tool to render it a comparatively inexpensive thread form.

1-5 LEAD-ANGLE SELECTION

Selecting the lead angle to provide a desired torque-to-axial-force ratio T/F is the central problem in power-screw design. Three versions of the power-screw problem can arise, and each may be treated separately.

Motion in a direction opposite to the direction of the applied force, as in raising a weight (Fig. 1-4a, b), involves a torque T and a force F which are related by the equation

$$\frac{T}{Fr_p} - \mu_c \frac{r_c}{r_p} = \tan(\rho + \lambda) \tag{1-11}$$

where μ_c = collar friction, and ρ = a thread-friction parameter.

If the screw or nut does not rotate under the force of the load alone, as in Fig. 1-4c, then the torque required to establish motion in the direction of the load, as in lowering a weight, is

$$\frac{T}{Fr_p} - \mu_c \frac{r_c}{r_p} = \tan(\rho - \lambda) \tag{1-12}$$

Finally, if the axial force of the load itself causes the screw to rotate and the load to translate when no external torque is applied, the screw is said to *overhaul,* and an external torque is required to restrain the motion. The magnitude of this torque is given by

$$-\frac{T}{Fr_p} - \mu_c \frac{r_c}{r_p} = \tan(\rho - \lambda) \tag{1-13}$$

The parameter ρ appearing in these equations is defined by

$$\tan \rho = \frac{\mu_t}{\cos \phi} \tag{1-14}$$

where
$$\phi = \tan^{-1}(\tan \alpha_1 \cos \lambda) \tag{1-15}$$

and α_1 is the leading flank angle.

Any one of Eqs. (1-11), (1-12), or (1-13) together with Eqs. (1-14) and (1-15) constitute a set of transcendental equations; the solution of these gives the lead angle corresponding to any desired torque-to-axial-force ratio.

1-6 EFFICIENCY, LUBRICATION, AND BACKLASH

The efficiency η is defined as

$$\eta = \frac{T_o}{T} \tag{1-16}$$

Notes ▪ Drawings ▪ Ideas

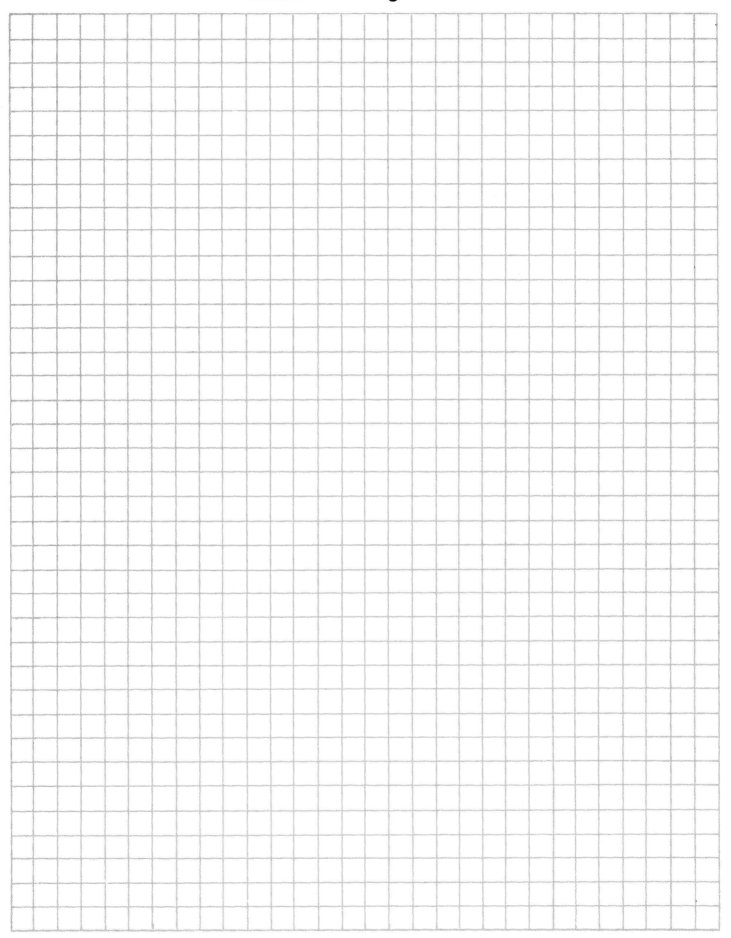

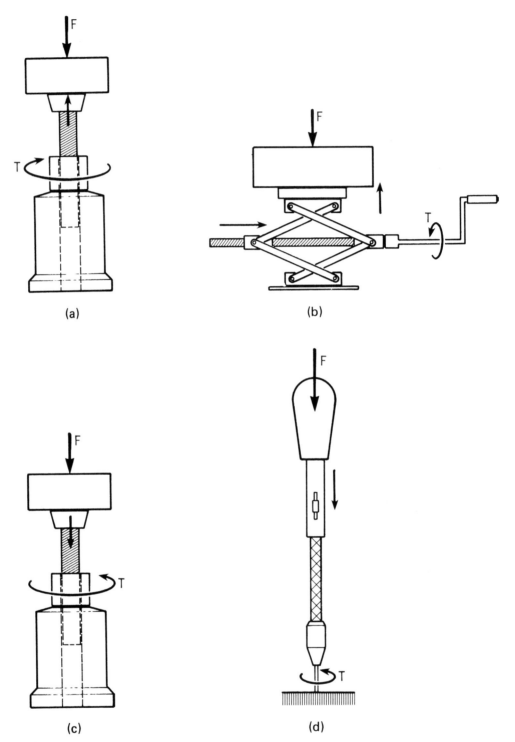

FIG. 1-4 (*a*) and (*b*) Motion opposes the applied force; (*c*) motion aids the applied force; (*d*) overhauling or back-driving.

where T_o is the torque that would be required to provide a given axial force if there were no friction. Hence

$$\eta = \frac{\cos \phi - \mu_c \tan \lambda}{\cos \phi + (\mu_t/\tan \lambda) + \mu_c(r_c/r_p)[(\cos \lambda/\tan \lambda) - \mu_t]} \qquad (1\text{-}17)$$

Figure 1-5 shows the dramatic effect of the coefficients of friction μ_t and μ_c on the efficiency of Acme threads when $r_c/r_p = 2.0$. If the collar friction is zero, the effect of thread friction on the screw efficiency is shown by curve A. Changing the collar friction from 0 to 0.10 causes the maximum possible efficiency with $\mu_t = 0$ to drop from 100 percent to slightly less than 21 percent.

Since r_c/r_p is usually greater than 1.0 for power screws, we see from comparing curves B and C, for $\mu_t = 0.10$, that in either case the reduction of μ_c to a small value by the use of adequate lubrication or by incorporating ball or tapered roller-thrust bearings is very important if high efficiency is to be obtained.

Unless μ_t is also small, however, the efficiency of the power screw may still be considerably less than 90 percent. Complete lubrication with a high-pressure lithium-soap base grease having good lubricating properties over the expected temperature range is therefore essential. Enclosure of the screw within a metal cylinder or a boot is recommended whenever there is the possibility of dirt and/or corrosive contaminants increasing the coefficient of friction.

Power screws used to provide axial force in a horizontal direction when rotating both clockwise and counterclockwise may display *backlash*, which means that when

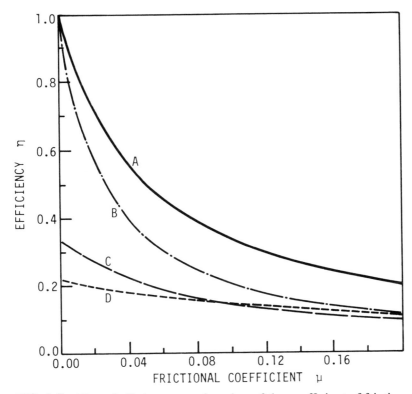

FIG. 1-5 Plot of efficiency as a function of the coefficient of friction for Acme threads with $r_c/r_p = 2.0$. A, $\mu_c = 0$, $\mu = \mu_t$; B, $\mu_t = 0$, $\mu = \mu_c$; C, $\mu_t = 0.10$, $\mu = \mu_c$; D, $\mu_c = 0.10$, $\mu = \mu_t$.

the screw changes its direction of rotation it must rotate through a small angle to change the bearing surface from one flank to the other. The space between the threads of the screw and of the nut, shown in Fig. 1-6, is a direct consequence of the manufacturing allowance and tolerance required to ensure assembly of mass-produced parts. Backlash may be greatly reduced by using a split nut which is usually adjustable to make up for wear.

1-7 BALL SCREWS

Ball screws, shown in Fig. 1-7, further reduce μ_t by replacing sliding friction with rolling friction. From Table 1-3 and Fig. 1-5 it follows that efficiencies on the order of 80 percent or greater are possible.

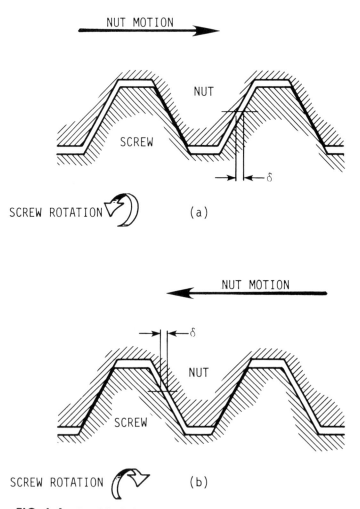

FIG. 1-6 Backlash in power screws. The distance δ is the sum of the tolerance and the allowance. (*a*) Position of screw and nut before change in rotation; (*b*) after change in rotation.

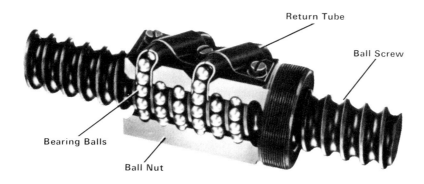

FIG. 1-7 Ball screw. *(Saginaw Steering Gear Division, General Motors Corporation.)*

Ball-screw life X in inches of travel and axial load F in pounds force are related by

$$F = F_b \left(\frac{10^6}{X}\right)^{1/3} \tag{1-18}$$

The *basic load rating* F_b is

$$F_b = 4500 Z^{2/3} d_i^{1.8} n_t^{0.86} L^{1/3} \tag{1-19}$$

where Z = the number of load-carrying balls per turn, d_i = the ball diameter in inches, n_t = the number of ball turns under a unidirectional load, and L = the lead in inches. In these units the *basic static-thrust capacity* T_b in pounds force is given by

$$T_b = 10^4 n_t Z d_i^2 \tag{1-20}$$

If SI units are used, the preceding relations become

$$F = F_b \left(\frac{25\,400}{X}\right)^{1/3} \tag{1-21}$$

where F and F_b are in newtons and X is in meters. Also,

$$F_b = 20.16 Z^{2/3} d_i^{1.8} n_t^{0.86} L^{1/3} \tag{1-22}$$

where d_i and L are in millimeters. In a similar manner, T_b in newtons is given by

$$T_b = 68.95 n_t Z d_i^2 \tag{1-23}$$

If the load on a ball screw varies in a known manner, then an equivalent load may be calculated from the equation

$$F = \left(\frac{1}{100} \sum_{i=1}^{m} C_i F_i\right)^{1/3} \tag{1-24}$$

Notes ▪ Drawings ▪ Ideas

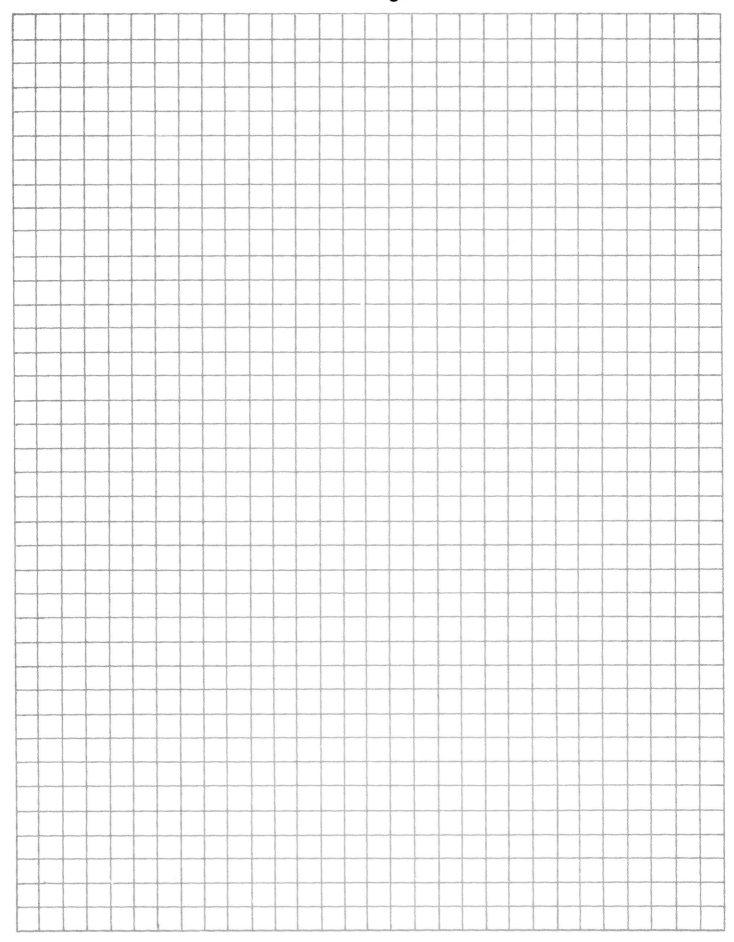

where C_i = the percentage of time that load F_i acts. Thus

$$\sum_{i=1}^{m} C_i = 100 \qquad (1\text{-}25)$$

The axial deflection may be calculated from formulas to be found in Appendix A-1 of Ref. [1-5].

Rather than tabulate the variables in these equations, most ball-screw manufacturer's catalogs simply tabulate F_b and T_b (see Table 1-4). To select a commercially available ball screw from such catalogs or tables, we calculate F from Eq. (1-24); if more than one load acts, we calculate X for the intended life of the ball screw and then find the required basic load rating from either Eq. (1-18) or (1-21) depending on the units used.

If no commercial ball screw has a basic load rating sufficiently close to the desired value of F_b, or if they are unsuited for other reasons, then Eqs. (1-19) and (1-20) or (1-22) and (1-23) would be used in the design of a custom ball screw. It is worth noting, however, that there are limitations on the use of these relations based on the geometry and materials used. Use the supplier's catalogs for details.

Realization of the rated life requires correct lubrication. Oil bath and/or oil mist are preferred, with the lubricant having a viscosity of approximately 140 Saybolt Universal seconds at the operating temperature recommended. Where oil lubrication is not possible, sodium-base, lithium, and lithium-soda greases are usually recommended.

TABLE 1-4 Sizes and Capacities of Ball-Bearing Lead Screws

Major diameter, in	Lead, in, **mm**	Ball diameter, in	Dynamic, capacity, lb	Static capacity, lb
0.750	0.200	0.125	1 242	4 595
	0.250	0.125	1 242	4 495
0.875	0.200	0.125	1 336	5 234
	0.250	0.125	1 336	5 234
1.000	0.200	0.125	1 418	5 973
	0.200†	0.156	1 909	7 469
	0.250	0.125	1 418	5 973
	0.250	0.156	1 909	7 469
	0.250	0.187	. . .	. . .
	0.400	0.125	1 418	5 973
	0.400	0.187	. . .	. . .
1.250	0.200	0.125	1 904	9 936
	0.200†	0.156	2 583	12 420
	0.250	0.125	1 904	9 936
	0.250	0.156	2 583	12 420
	0.250	0.187	3 304	15 886
1.500	0.200	0.125	2 046	11 908
	0.200†	0.156	2 786	14 881
	0.250	0.156	2 786	14 881
	0.250	0.187	3 583	18 748
	0.500	0.156	2 786	14 881
	0.500	0.250	5 290	24 762

TABLE 1-4 Sizes and Capacities of Ball-Bearing Lead Screws (*Continued*)

Major diameter, in	Lead, in, **mm**	Ball diameter, in	Dynamic, capacity, lb	Static capacity, lb
1.500	**5**†	0.125	2 046	11 908
	5	0.156	2 787	14 881
	10	0.156	2 786	14 881
	10	0.250	5 290	24 762
	10	0.312	7 050	29 324
1.750	0.200	0.125	2 179	13 879
	0.200†	0.156	2 968	17 341
	0.250	0.156	2 968	17 341
	0.250	0.187	3 829	20 822
	0.500	0.187	3 829	20 882
	0.500	0.250	5 664	27 917
	0.500	0.312	7 633	33 232
2.000	0.200	0.125	2 311	15 851
	0.200†	0.156	3 169	19 801
	0.250	0.156	3 169	19 801
	0.250	0.187	4 033	23 172
	0.400	0.250	6 043	31 850
	0.500	0.312	8 135	39 854
	5	0.125	2 311	15 851
	5†	0.156	3 169	19 801
	6	0.156	3 169	19 801
	6	0.187	4 033	23 172
	10	0.250	6 043	31 850
	10	0.312	8 135	39 854
2.250	0.250	0.156	3 306	22 262
	0.250	0.187	4 266	26 684
	0.500	0.312	8 593	44 780
	0.500	0.375	10 862	53 660
2.500	0.200	0.125	2 511	19 794
	0.200	0.156	3 134	24 436
	0.250	0.187	4 410	29 671
	0.400	0.250	6 633	39 746
	0.500	0.312	9 015	49 701
	0.500	0.375	10 367	59 308
	5	0.125	2 511	19 794
	5†	0.156	3 134	24 436
	10	0.250	6 633	39 746
	10	0.312	9 015	49 701
3.000	0.250	0.187	4 810	35 570
	0.400	0.250	7 125	47 632
	0.500	0.375	12 560	71 685
	0.660	0.375	12 560	71 685
	10	0.250	7 125	47 632
	10	0.312	9 744	58 648
3.500	0.500	0.312	10 360	69 287
	0.500	0.375	13 377	83 514
	1.000	0.500	19 812	111 510
	1.000	0.625	26 752	139 585
4.000	0.500	0.375	14 088	95 343
	1.000	0.500	21 066	127 282

†These values are not recommended; consult manufacturer.
SOURCE: From 20th Century Machine Company, Sterling Heights, Michigan, by permission.

High efficiency for ball screws is obtained at the expense of stricter lubrication requirements, longer nuts to carry equal axial loads, and the characteristic of overhauling, or *backdriving,* under comparatively light loads, often no more than the weight of the nut itself.

Two ball nuts may be drawn together to reduce backlash; this is similar to the split-nut mechanism used for friction power screws. Spring loading can also be provided.

1-8 REVERSING SCREWS

Reciprocating motion may be obtained from steady rotational motion by means of a reversing screw similar to that shown in Fig. 1-8. A very common example is to be found on the level-wind fisherman's casting reel. Lead angles of such screws are generally limited to a range from 15 to 30 degrees for reversing screws in which screw rotation drives the carrier or cage assembly. This is to limit the torque applied to the cage assembly due to friction from the screw. Commercially available reversing screws have been produced with stroke lengths from 11.2 mm to 2.473 m.

Similar reversing screws, but without turnarounds and ball nuts, are used for push drills and screwdrivers in which backdriving is intended. They use a square-thread profile and their lead angles may be between 50 and 80 degrees.

1-9 ROLLER SCREWS

Intermediate values of thread friction between those of ball screws and conventional screws (sliding friction, face contact) may be obtained from roller screws (Fig. 1-9a), in which the nut is internally threaded with the same thread profile, lead angle, and number of threads as the central screw. In the planetary roller design shown in Fig. 1-9b the rollers are also threaded so they may roll between the central screw and the nut and thus advance the nut. Gear teeth are cut in the roller ends which mesh with internal gears at each end of the nut to ensure that the rollers always roll on the nut without sliding. This prevents the rollers from moving out of the nut, as they would if they slide along the internal threads in the nut.

The recirculating roller design shown in Fig. 1-9b differs from the planetary roller design in that the threads on the rollers are replaced by circumferential grooves

FIG. 1-8 Ball reversing screw. *(NORCO, Inc.)*

20 GEARING: A MECHANICAL DESIGNERS' WORKBOOK

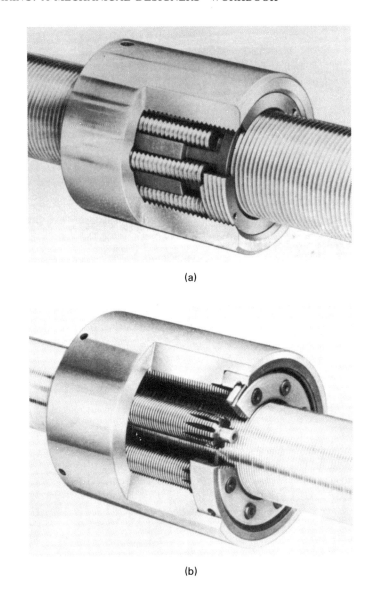

FIG. 1-9 (*a*) Recirculating roller screw; (*b*) planetary roller screw. *(LTI-Transrol.)*

with the same profiles as the threads on the nut and central screw, but with zero lead. Although separated by a cage, they otherwise roll freely between the nut and central screw and advance faster than the nut; i.e., after making a complete circuit about the inside of the nut, each roller will have advanced relative to the nut by the amount of the thread lead. This advance relative to the nut is removed at the end of each circuit by guiding the roller into a slot in the roller cage where it is moved back to its starting position relative to the nut.

Recirculating roller screws have a smaller load capacity than the planetary roller screws for a given size of nut and screw, but they may be positioned more accurately because they can be produced with smaller lead angles. No provision has been made in the existing designs for preloading to remove backlash, as has been done for ball screws.

Notes · Drawings · Ideas

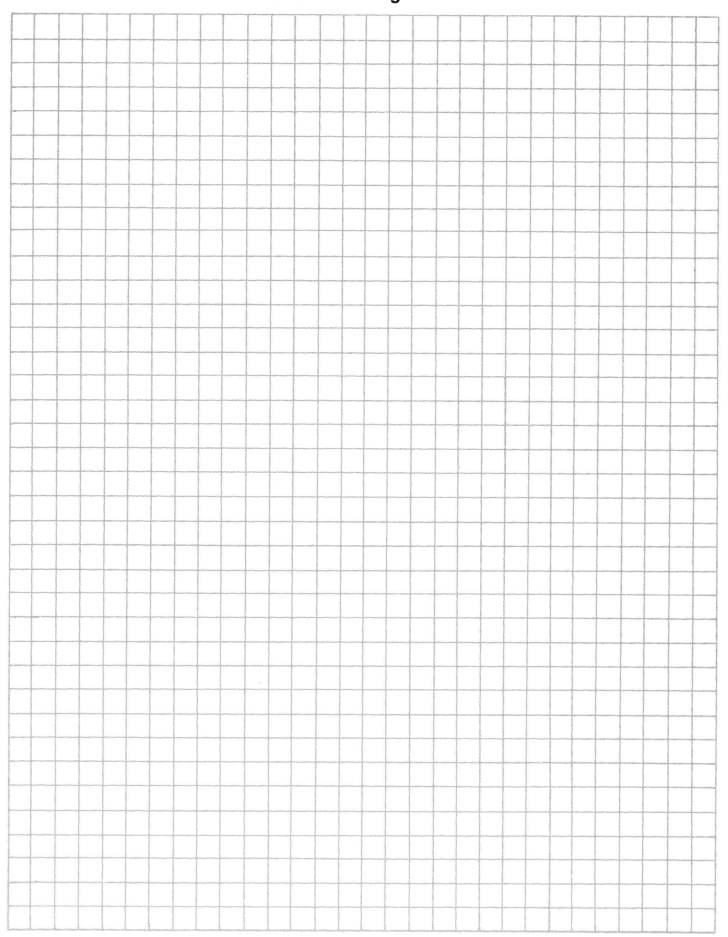

REFERENCES

1-1 "Screw Threads, Nomenclature, Definitions, and Letter Symbols," ANSI B1.7-1977, American Society of Mechanical Engineers, New York, 1977.

1-2 "Acme Screw Threads," ANSI B1.5-1977, American Society of Mechanical Engineers, New York, 1977.

1-3 "Stub Acme Screw Threads," ANSI B1.8-1977, American Society of Mechanical Engineers, New York, 1977.

1-4 "Buttress Screw Threads," ANSI B1.9-1973 (R1979), American Society of Mechanical Engineers, New York, 1973.

1-5 "Ball Screws," ANSI B5.48-1977, American Society of Mechanical Engineers, New York, 1977.

chapter 2
SPUR GEARS

JOSEPH E. SHIGLEY
Professor Emeritus
The University of Michigan
Ann Arbor, Michigan

2-1 DEFINITIONS

Spur gears are used to transmit rotary motion between parallel shafts. They are cylindrical, and the teeth are straight and parallel to the axis of rotation.

The *pinion* is the smaller of two mating gears; the larger is called the *gear* or the *wheel*.

The *pitch circle, B* in Fig. 2-1, is a theoretical circle upon which all calculations are based. The *operating pitch circles* of a pair of gears in mesh are tangent to each other.

The *circular pitch, p* in Fig. 2-1, is the distance, measured on the theoretical pitch circle, from a point on one tooth to a corresponding point on an adjacent tooth. The circular pitch is measured in inches or in millimeters. Note, in Fig. 2-1, that the circular pitch is the sum of the *tooth thickness t* and the *width of space*.

The *pitch diameter, d* for the pinion and *D* for the gear, is the diameter of the pitch circle; it is measured in inches or in millimeters.

The *module m* is the ratio of the theoretical pitch diameter to the number of teeth N. The module is the metric index of tooth sizes and is always given in millimeters.

The *diametral pitch* P_d is the ratio of the number of teeth on a gear to the theoretical pitch diameter. It is the index of tooth size when U.S. customary units are used and is expressed as teeth per inch.

The *addendum a* is the radial distance between the top land *F* and the pitch circle *B* in Fig. 2-1. The *dedendum b* is the radial distance between the pitch circle *B* and the dedendum or *root circle D* in Fig. 2-1. The *whole depth h*, is the sum of the addendum and dedendum.

The clearance circle *C* in Fig. 2-1 is tangent to the addendum circle of the mating gear. The distance from the clearance circle to the bottom land is called the *clearance c*.

Backlash is the amount by which the width of a tooth space exceeds the thickness of the engaging tooth measured on the pitch circle.

Undercutting (see distance *u* in Fig. 2-1) occurs under certain conditions when a small number of teeth are used in cutting a gear.

Table 2-1 lists all the relations described above. Additional terminology is shown in Fig. 2-2. Here line *OP* is the *line of centers* connecting the rotation axes of a pair of meshing gears. Line *E* is the *pressure line*, and the angle ϕ is the *pressure angle*. The resultant force vector between a pair of operating gears acts along this line.

The pressure line is tangent to both *base circles C* at points *F*. The operating diameters of the pitch circles depend on the center distance used in mounting the gears, but the base circle diameters are constant and depend only on how the tooth forms were generated, because they form the *base* or the starting point of the involute profile. See Ref. [2-1] or [2-2].

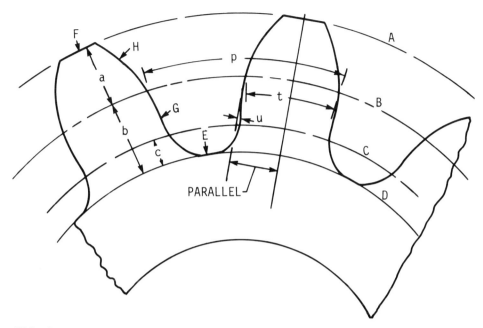

FIG. 2-1 Terminology of gear teeth. A, addendum circle; B, pitch circle; C, clearance circle; D, dedendum circle; E, bottom land; F, top land; G flank; H, face; a = addendum distance; b = dedendum distance; c = clearance distance; p = circular pitch; t = tooth thickness; u = undercut distance.

Line aPb is the *line of action*. Point a is the *initial point of contact*. This point is located at the intersection of the addendum circle of the gear with the pressure line. Should point a occur on the other side of point F on the pinion base circle, the pinion flank would be *undercut* during generation of the profile.

Point b of Fig 2-2 is the *final point of contact*. This point is located at the intersection of the addendum circle of the pinion with the pressure line. For no under-

TABLE 2-1 Basic Formulas for Spur Gears

Quantity desired	Formula	Equation number
Diametral pitch P_d	$P_d = \dfrac{N}{d}$	(2-1)
Module m	$m = \dfrac{d}{N}$	(2-2)
Circular pitch p	$p = \dfrac{\pi d}{N} = \pi m$	(2-3)
Pitch diameter, d or D	$d = \dfrac{N}{P_d} = mN$	(2-4)

26 GEARING: A MECHANICAL DESIGNERS' WORKBOOK

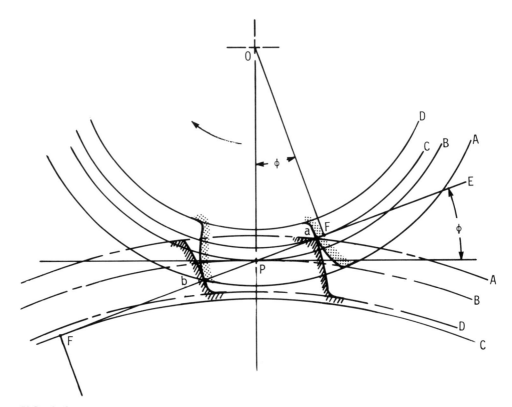

FIG. 2-2 Layout drawing of a pair of spur gears in mesh. The pinion is the driver and rotates clockwise about axis at O. A, addendum circles; B, pitch circles; C, base circles; D, dedendum circles; E, pressure line; F, tangent points; P, pitch point; a, initial point of contact; b, final point of contact.

cutting of the gear teeth, point b must be located between the pitch point P and point F on the base circle of the gear.

Line aP represents the *approach* phase of tooth contact; line Pb is the *recess* phase. Tooth contact is a sliding contact throughout the line of action except for an instant at P when contact is pure rolling. The nature of the sliding is quite different during the approach action than for the recess action; and bevel-gear teeth, for example, are generated to obtain more recess action, thus reducing wear.

TABLE 2-2 Standard and Commonly Used Tooth Systems for Spur Gears

Tooth system	Pressure angle ϕ, deg	Addendum a	Dedendum b
Full depth	20	$1/P_d$ or $1m$	$1.25/P_d$ or $1.25m$ $1.35/P_d$ or $1.35m$
	$22\frac{1}{2}$	$1/P_d$ or $1m$	$1.25/P_d$ or $1.25m$ $1.35/P_d$ or $1.35m$
	25	$1/P_d$ or $1m$	$1.25/P_d$ or $1.25m$ $1.35/P_d$ or $1.35m$
Stub	20	$0.8/P_d$ or $0.8m$	$1/P_d$ or $1m$

TABLE 2-3 Diametral Pitches in General Use

Coarse pitch   2, 2¼, 2½, 3, 4, 6, 8, 10, 12, 16
Fine pitch   20, 24, 32, 40, 48, 64, 96, 120, 150, 200

Instead of using the theoretical pitch circle as an index of tooth size, the base circle, which is a more fundamental distance, can be used. The result is called the *base pitch* p_b. It is related to the circular pitch p by the equation

$$p_b = p \cos \phi \tag{2-5}$$

If, in Fig. 2-2, the distance from a to b exactly equals the base pitch, then, when one pair of teeth are just beginning contact at a, the preceding pair will be leaving contact at b. Thus, for this special condition, there is never more or less than one pair of teeth in contact. If the distance ab is greater than the base pitch but less than twice as much, then when a pair of teeth come into contact at a, another pair of teeth will still be in contact somewhere along the line of action ab. Because of the nature of this tooth action, usually one or two pairs of teeth in contact, a useful criterion of tooth action, called the *contact ratio* m_c, can be defined. The formula is

$$m_c = \frac{L_{ab}}{p_b} \tag{2-6}$$

where L_{ab} = distance ab, the length of the line of action. Do not confuse the contact ratio m_c with the module m.

2-2 TOOTH DIMENSIONS AND STANDARDS

The American Gear Manufacturer's Association (AGMA) publishes much valuable reference data.† The details on nomenclature, definitions, and tooth proportions can be found in AGMA 112.04 and 201.02 for spur gears. Table 2-2 contains the most used tooth proportions. The hob tip radius r_f varies with different cutters; $0.300/P_d$ or $0.300m$ is the usual value. Tables 2-3 and 2-4 list the modules and pitches in general use. Cutting tools can be obtained for all these sizes.

2-3 FORCE ANALYSIS

In Fig. 2-3 a gear, not shown, exerts force W against the pinion at pitch point P. This force is resolved into two components, a radial force W_r, acting to separate the gears, and a tangential component W_t, which is called the *transmitted load*.

Equal and opposite to force W is the shaft reaction F, also shown in Fig. 2-3. Force F and torque T are exerted by the shaft on the pinion. Note that torque T opposes the force couple made up of W_t and F_x separated by the distance $d/2$. Thus

$$T = \frac{W_t d}{2} \tag{2-7}$$

---

†See Chap. 4 for a special note on AGMA.

TABLE 2-4 Modules in General Use

Preferred	1, 1.25, 1.5, 2, 2.5, 3, 4, 5, 6, 8, 10, 12, 16, 20, 25, 32, 40, 50
Next choice	1.125, 1.375, 1.75, 2.25, 2.75, 3.5, 4.5, 5.5, 7, 9, 11, 14, 18, 22, 28, 36, 45

where T = torque, lb·in (N·m)
W_t = transmitted load, lb (N)
d = operating pitch diameter, in (m)

The *pitch-line velocity* v is given by

$$v = \frac{\pi d n_P}{12} \text{ ft/min} \qquad v = \frac{\pi d n_P}{60} \text{ m/s} \qquad (2\text{-}8)$$

where n_P = pinion speed in revolutions per minute (r/min). The power transmitted is

$$P = \begin{cases} \dfrac{W_t v}{33\,000} & \text{hp} \\ W_t v & \text{kW} \end{cases} \qquad (2\text{-}9)$$

2-4 FUNDAMENTAL AGMA RATING FORMULAS†

Many of the terms in the formulas that follow require lengthy discussions and considerable space to list their values. This material is considered at length in Chap. 4 and so is omitted here.

2-4-1 Pitting Resistance

The basic formula for *pitting resistance,* or *surface durability,* of gear teeth is

$$s_c = C_p \left(\frac{W_t C_a}{C_v} \frac{C_s}{dF} \frac{C_m C_f}{I} \right)^{1/2} \qquad (2\text{-}10)$$

where s_c = contact stress number, lb/in² (MPa)
C_p = elastic coefficient, (lb/in²)$^{1/2}$ [(MPa)$^{1/2}$]; see Eq. (4-77) and Table 4-4
W_t = transmitted tangential load, lb (N)
C_a = application factor for pitting resistance; see Table 4-3
C_s = size factor for pitting resistance; use 1.0 or more until values are established
C_m = load distribution factor for pitting resistance; use Tables 2-5 and 2-6
C_f = surface condition factor; use 1.0 or more until values are established
C_v = dynamic factor for pitting resistance; use Fig. 4-4; multiply v in meters per second by 197 to get feet per minute

†See Ref. [4-1].

Notes · Drawings · Ideas

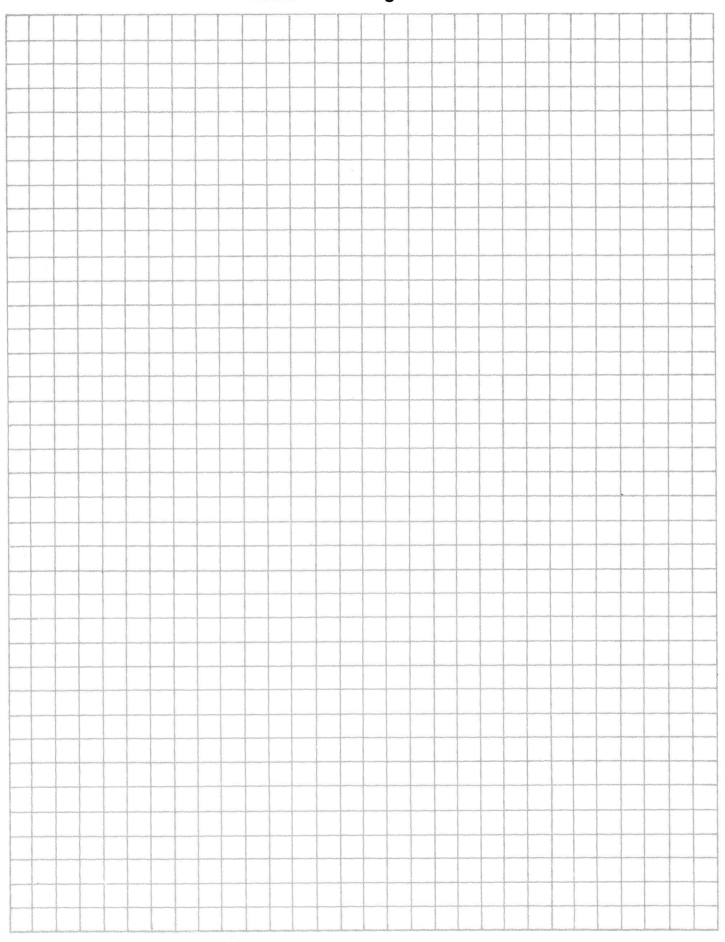

30 GEARING: A MECHANICAL DESIGNERS' WORKBOOK

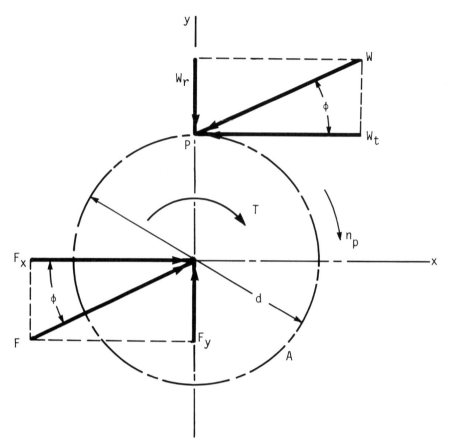

FIG. 2-3 Force analysis of a pinion. A, operating pitch circle; d, operating pitch diameter; n_p, pinion speed; ϕ, pressure angle; W_t, transmitted tangential load; W_r, radial tooth load; W, resultant tooth load; T, torque; F, shaft force reaction.

$\quad d =$ operating pitch diameter of pinion, in (mm)
$\quad\;\; = 2C/(m_G + 1.0)$ for external gears
$\quad\;\; = 2C/(m_G - 1.0)$ for internal gears
$\quad C =$ operating center distance, in (mm)
$m_G =$ gear ratio (never less than 1.0)
$\quad F =$ net face width of narrowest member, in (mm)
$\quad I =$ geometry factor for pitting resistance; use Eq. (4-24) with $C_\psi = 1.0$

Allowable Contact Stress Number. The contact stress number s_c, used in Eq. (2-10), is obtained from the *allowable contact stress number* s_{ac} by making several adjustments as follows:

$$s_c \leq s_{ac} \frac{C_L C_H}{C_T C_R} \tag{2-11}$$

where $s_{ac} =$ allowable contact stress number, lb/in^2 (MPa); see Fig. 4-40
$\quad\; C_L =$ life factor for pitting resistance; use Fig. 4-49
$\quad C_H =$ hardness ratio factor; use Figs. 4-47 and 4-48

TABLE 2-5 Load-Distribution Factors C_m and K_m for Spur Gears Having a Face Width of 6 in (150 mm) and Greater†

Face-diameter ratio F/d	Contact	C_m, K_m
1 or less	95% face width contact at one-third torque	1.4 at one-third torque
	95% face width contact at full torque	1.1 at full torque
	75% face width contact at one-third torque	1.8 at one-third torque
	95% face width contact at full torque	1.3 at full torque
	35% face width contact at one-third torque	3.0 at one-third torque
	95% face width contact at full torque	1.9 at full torque
	20% face width contact at one-third torque	5.0 at one-third torque
	75% face width contact at full torque	2.5 at full torque
	Teeth are crowned: 35% face width contact at one-third torque	2.5 at one-third torque
	85% face width contact at full torque	1.7 at full torque
Over 1 and less than 2	Calculated combined twist and bending of pinion not over 0.001 in (0.025 mm) over entire face: Pinion not over 250 bhn hardness: 75% face width contact at one-third torque	2.0 at one-third torque
	95% face width contact at full torque	1.4 at full torque
	30% face width contact at one-third torque	4.0 at one-third torque
	75% face width contact at full torque	3.0 at full torque

†For an alternate approach see Eq. (4-21).
SOURCE: AGMA 215.01 and 225.01.

C_T = temperature factor for pitting resistance; use 1.0 or more, but see Sec. 4-5-1

C_R = reliability factor for pitting resistance; use Table 4-6

Pitting Resistance Power Rating. The allowable power rating P_{ac} for pitting resistance is given by

TABLE 2-6 Load-Distribution Factors C_m and K_m for Spur Gears

Condition of support	Face width			
	Up to 2 in (50 mm)	6 in (150 mm)	9 in (225 mm)	Over 16 in (400 mm)
Accurate mounting, low bearing clearances, minimum elastic deflection, precision gears	1.3	1.4	1.5	1.8
Less rigid mountings, less accurate gears, contact across full face	1.6	1.7	1.8	2.0
Accuracy and mounting such that less than full-face contact exists	Over 2.0			

SOURCE: AGMA 215.01 and 225.01. For an alternate approach see Eq. (4-21).

$$P_{ac} = \begin{cases} \dfrac{n_P F}{126\,000} \dfrac{IC_v}{C_s C_m C_f C_a} \left(\dfrac{ds_{ac}}{C_p} \dfrac{C_L C_H}{C_T C_R} \right)^2 & \text{hp} \\ \dfrac{n_P F}{1.91(10^7)} \dfrac{IC_v}{C_s C_m C_f C_a} \left(\dfrac{ds_{ac}}{C_p} \dfrac{C_L C_H}{C_T C_R} \right)^2 & \text{kW} \end{cases} \quad (2\text{-}12)$$

2-4-2 Bending Strength

The basic formula for the bending stress number in a gear tooth is

$$S_t = \begin{cases} \dfrac{W_t K_a}{K_v} \dfrac{P_d}{F} \dfrac{K_s K_m}{J} & \text{lb/in}^2 \\ \dfrac{W_t K_a}{K_v} \dfrac{1.0}{Fm} \dfrac{K_s K_m}{J} & \text{MPa} \end{cases} \quad (2\text{-}13)$$

where s_t = bending stress number, lb/in² (MPa)
K_a = application factor for bending strength; use Table 4-3
K_s = size factor for bending strength; use 1.0 or more until values are established
K_m = load-distribution factor for bending strength; use Tables 2-5 and 2-6
K_v = dynamic factor for bending strength; use Fig. 4-4; multiply v in meters per second by 197 to get feet per minute
J = geometry factor for bending strength; use Eq. (4-46) with $C_\psi = 1.0$ and Figs. 4-11 to 4-22
m = module, mm
P_d = nominal diametral pitch, teeth per inch

Notes · Drawings · Ideas

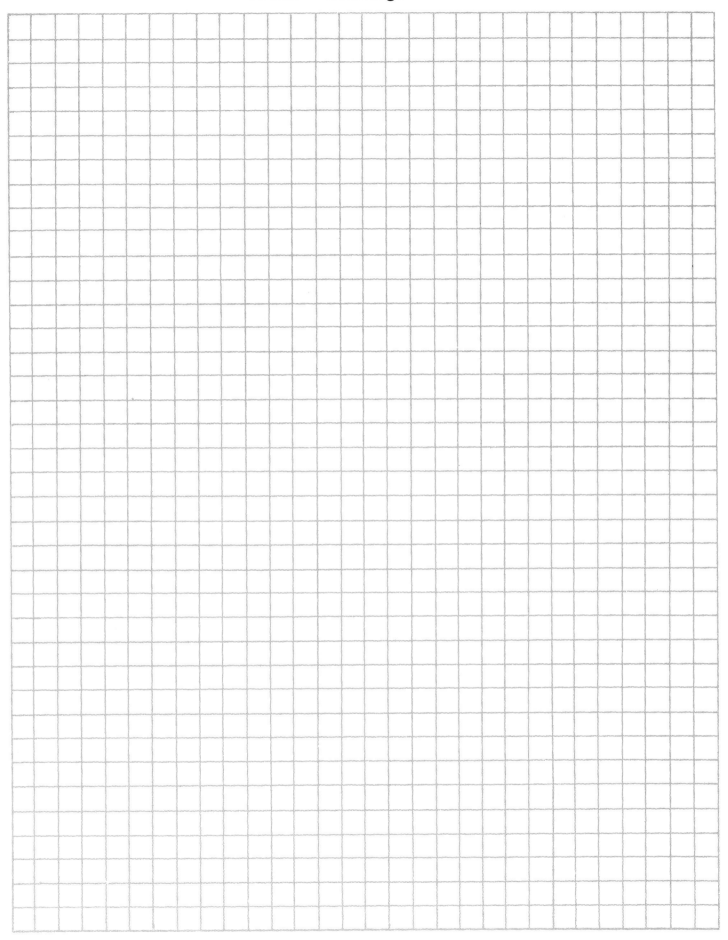

Allowable Bending Stress Number. The bending stress number s_t in Eq. (2-13) is related to the *allowable bending stress number* s_{at} by

$$s_t \leq \frac{s_{at} K_L}{K_T K_R} \qquad (2\text{-}14)$$

where s_{at} = allowable bending stress number, lb/in² (MPa); use Fig. 4-41
K_L = life factor for bending strength; use Figs. 4-49 and 4-50
K_T = temperature factor for bending strength; use 1.0 or more; see Sec. 4-5-1
K_R = reliability for bending strength; use Table 4-6

Bending Strength Power Rating. The allowable power rating P_{at} for bending strength is given by

$$P_{at} = \begin{cases} \dfrac{n_P d K_v}{126\,000 K_a} \dfrac{FJ}{P_d K_s K_m} \dfrac{s_{at} K_L}{K_R K_T} & \text{hp} \\ \dfrac{n_P d K_v}{1.91(10)^7 K_a} Fm \dfrac{J}{K_s K_m} \dfrac{s_{at} K_L}{K_R K_T} & \text{kW} \end{cases} \qquad (2\text{-}15)$$

REFERENCES

2-1 Joseph E. Shigley and John J. Uicker, Jr., *Theory of Machines and Mechanisms,* McGraw-Hill, 1980.

2-2 Joseph E. Shigley, *Mechanical Engineering Design,* first metric edition, McGraw-Hill, 1986.

chapter 3
BEVEL AND HYPOID GEARS

THEODORE J. KRENZER, M.S.
Manager, Gear Theory Department
Gleason Machine Division
Rochester, New York

ROBERT G. HOTCHKISS, B.S.
Director, Gear Technology
Gleason Machine Division
Rochester, New York

3-1 INTRODUCTION

This chapter provides you with information necessary to design a bevel or hypoid-gear set. It includes guidelines for selecting the type and size of a gear set to suit the application requirements. Equations and graphs are provided for calculating gear-tooth geometry, strength, surface durability, and bearing loads.

Although the text provides sufficient data to design a gear set, reference is also made to appropriate American Gear Manufacturer's Association (AGMA) publications and software available for computer-aided design.

3-2 TERMINOLOGY

3-2-1 Types of Bevel and Hypoid Gears

Straight-bevel gears are the simplest form of bevel gears. The teeth are straight and tapered, and if extended inward, they would pass through the point of intersection of the axes. See Fig. 3-1.

Spiral-bevel gears have teeth that are curved and oblique to their axes. The contact begins at one end of the tooth and progresses to the other. See Fig. 3-2.

Zerol bevel gears have teeth that are in the same general direction as straight-bevel gears and are curved similarly to spiral-bevel gears. See Fig. 3-3.

Hypoid gears are similar in appearance to spiral-bevel gears. They differ from spiral-bevel gears in that the axis of the pinion is offset from the axis of the gear. See Fig. 3-4.

3-2-2 Tooth Geometry

The nomenclature used in this chapter relative to bevel and hypoid gears is illustrated in Figs. 3-5, 3-6, and 3-7.

The following terms are used to define the geometry:

Addendum of pinion (gear) a_p (a_G) is the height that the tooth projects above the pitch cone.

FIG. 3-1 Straight-bevel set. *(Gleason Machine Division.)*

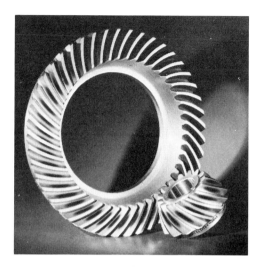

FIG. 3-2 Spiral-bevel set. *(Gleason Machine Division.)*

Backlash allowance B is the amount by which the circular tooth thicknesses are reduced to provide the necessary backlash in assembly.

Clearance c is the amount by which the dedendum in a given gear exceeds the addendum of its mating gear.

Cone distance, mean, A_m is the distance from the apex of the pitch cone to the middle of the face width.

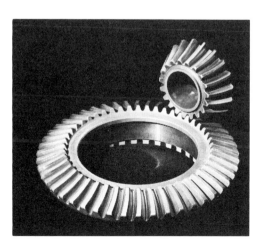

FIG. 3-3 Zerol bevel set. *(Gleason Machine Division.)*

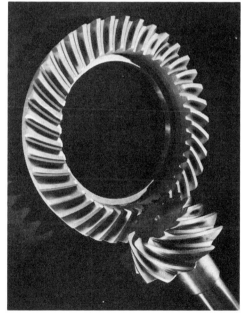

FIG. 3-4 Hypoid set. *(Gleason Machine Division.)*

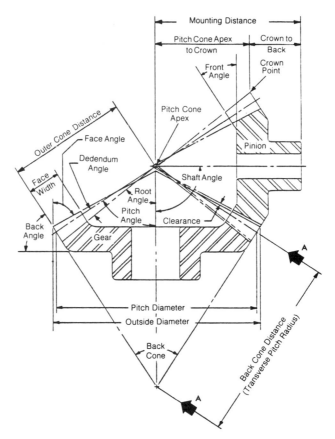

FIG. 3-5 Bevel-gear nomenclature—axial plane. Section *AA* is illustrated in Fig. 3-6.

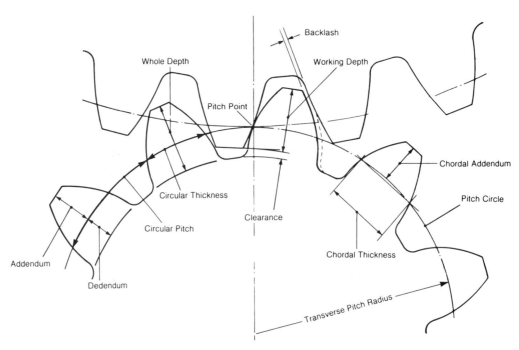

FIG. 3-6 Bevel-gear nomenclature—mean transverse section *AA* of Fig. 3-5.

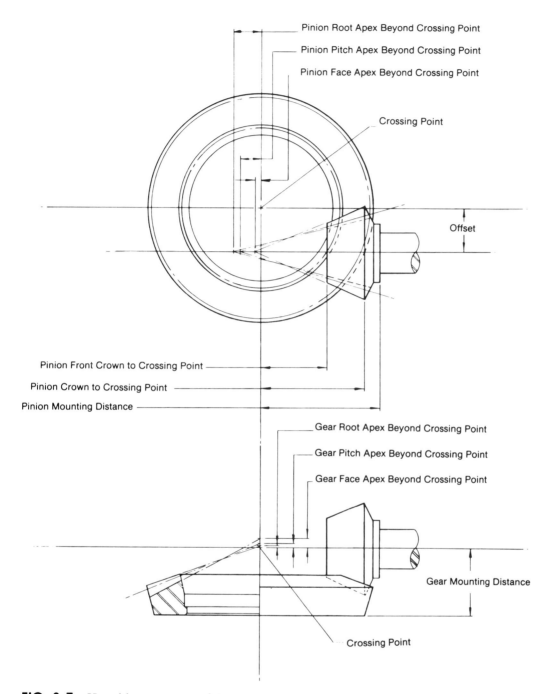

FIG. 3-7 Hypoid-gear nomenclature.

Cone distance, outer, A_o is the distance from the apex of the pitch cone to the outer ends of the teeth.

Control gear is the adopted term for bevel gearing in place of the term *master gear*, which implies a gear with all tooth specifications held to close tolerances.

Crown to crossing point on the pinion (gear) x_o (X_o) is the distance in an axial section from the crown to the crossing point, measured in an axial direction.

Cutter radius r_c is the nominal radius of the face-type cutter or cup-shaped grinding wheel that is used to cut or grind the spiral-bevel teeth.

Dedendum angle of pinion (gear) δ_P (δ_G) is the angle between elements of the root cone and pitch cone.

Dedendum angles, sum of, $\Sigma\delta$ is the sum of the pinion and gear dedendum angles.

Dedendum of pinion (gear) b_p (b_G) is the depth of the tooth space below the pitch cone.

Depth, mean whole, h_m is the tooth depth at midface.

Depth, mean working, h is the depth of engagement of two gears at midface.

Diametral pitch P_d is the number of gear teeth per unit of pitch diameter.

Face angle of pinion (gear) blank γ_o (Γ_o) is the angle between an element of the face cone and its axis.

Face apex beyond crossing point on the pinion (gear) G_o (Z_o) is the distance between the face apex and the crossing point on a bevel or hypoid set.

Face width F is the length of the teeth measured along a pitch-cone element.

Factor, mean addendum, c_1 is the addendum modification factor.

Front crown to crossing point on the pinion (gear) x_i (X_i) is the distance in an axial section from the front crown to the crossing point, measured in the axial direction.

Hypoid offset E is the distance between two parallel planes, one containing the gear axis and the other containing the pinion axis of a hypoid-gear set.

Number of teeth in pinion (gear) n (N) is the number of teeth contained in the whole circumference of the pitch cone.

Pitch angle of pinion (gear) γ (Γ) is the angle between an element of the pitch cone and its axis.

Pitch apex beyond crossing point on the pinion (gear) G (Z) is the distance between the pitch apex and the crossing point on a hypoid set.

Pitch diameter of pinion (gear) d (D) is the diameter of the pitch cone at the outside of the blank.

Pitch, mean circular, p_m is the distance along the pitch circle at the mean cone distance between corresponding profiles of adjacent teeth.

Pressure angle ϕ is the angle at the pitch point between the line of pressure which is normal to the tooth surface and the plane tangent to the pitch surface. It is specified at the mean cone distance.

Ratio, gear, m_G is the ratio of the number of gear teeth to the number of pinion teeth.

Root angle of pinion (gear) γ_R (Γ_R) is the angle between an element of the root cone and its axis.

Root apex beyond crossing point on the pinion (gear) G_R (Z_R) is the distance between the root apex and the crossing point on a bevel or hypoid set.

Shaft angle Σ is the angle between the axes of the pinion shaft and the gear shaft.

Spiral angle ψ is the angle between the tooth trace and an element of the pitch cone. It is specified at the mean cone distance.

Spiral-bevel gear, left-hand, is one in which the outer half of a tooth is inclined in the counterclockwise direction from the axial plane through the midpoint of the tooth, as viewed by an observer looking at the face of the gear.

Spiral-bevel gear, right-hand, is one in which the outer half of a tooth is inclined in the clockwise direction from the axial plane through the midpoint of the tooth, as viewed by an observer looking at the face of the gear.

Tangential force W_t is the force applied to a gear tooth at the mean cone distance in a direction tangent to the pitch cone and normal to a pitch-cone element.

Thickness of pinion (gear), mean circular, t (T) is the length of arc on the pitch cone between the two sides of the tooth at the mean cone distance.

Thickness of pinion (gear), mean normal chordal, t_{nc} (T_{nc}) is the chordal thickness of the pinion tooth at the mean cone distance in a plane normal to the tooth trace.

3-2-3 Calculation Methods

Four methods of blank design are commonly used in the design of bevel and hypoid gears:

1. Standard taper
2. Duplex taper
3. Uniform taper
4. Tilted root-line taper

The taper you select depends in some instances on the manufacturing equipment available for producing the gear set. Therefore, before starting calculations, you should familiarize yourself with the equipment and method used by the gear manufacturer.

3-3 GEAR MANUFACTURING

3-3-1 Methods of Generation

Generation is the basic process in the manufacture of bevel and hypoid gears in that at least one member of every set must be generated. The theory of generation as applied to these gears involves an imaginary generating gear which can be a crown gear, a mating gear, or some other bevel or hypoid gear. The gear blank or workpiece is positioned so that when it is rolled with the generating gear, the teeth of the workpiece are enveloped by the teeth of the generating gear.

In the actual production of the gear teeth, at least one tooth of the generating gear is described by the motion of the cutting tool or grinding wheel. The tool and its motion are carried on a rotatable machine member called a cradle, the axis of which is identical with the axis of the generating gear. The cradle and the workpiece roll together on their respective axes exactly as would the workpiece and the generating gear.

The lengthwise tooth curve of the generating gear is selected so that it is easily followed with a practical cutting tool and mechanical motion. Figure 3-8 illustrates the representation of a generating gear by a face-mill cutter. Figure 3-9 shows the basic machine elements of a bevel-gear face-mill generator.

Most generating gears are based on one of two fundamental concepts. The first is complementary crown gears, where two gears with 90° pitch angles fit together like mold castings. Each of the crown gears is the generating gear for one member of the

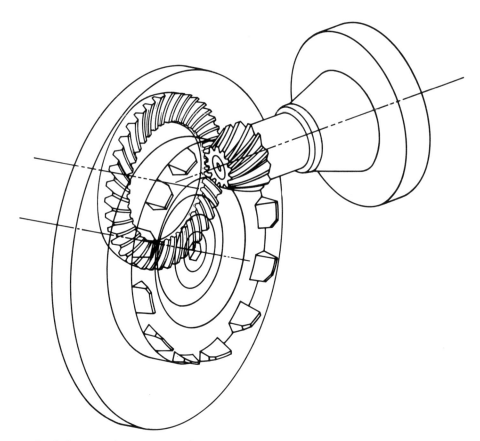

FIG. 3-8 Imaginary generating gear.

mating set. Gears generated in this manner have line contact and are said to be *conjugate* to each other. With the second concept, the teeth of one member are form-cut without generation. This member becomes the generating gear for producing the mating member. Again, gears generated in this manner are conjugate to each other.

3-3-2 Localization of Contact

Any displacement in the nominal running position of either member of a mating conjugate gear set shifts the contact to the edges of the tooth. The result is concentrated loading and irregular motion. To accommodate assembly tolerances and deflections resulting from load, tooth surfaces are relieved in both the lengthwise and profile directions. The resulting localization of the contact pattern is achieved by using a generating setup which is deliberately modified from the conjugate generating gear.

3-3-3 Testing

The smoothness and quietness of operation, the tooth contact pattern, the tooth size, the surface finish, and appreciable runout can be checked in a running test. This is a subjective test. The machine consists of two spindles that can be set at the correct shaft angle, mounting distances, and offset. The gear to be inspected is mounted on

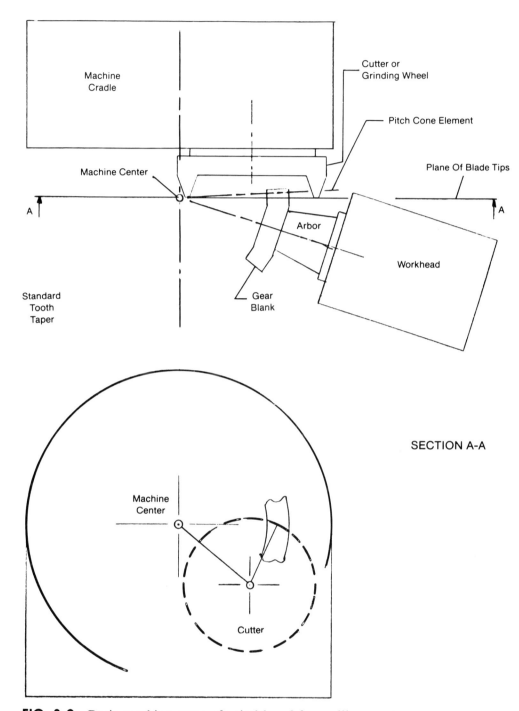

FIG. 3-9 Basic machine setup of spiral-bevel face-mill generator.

one spindle, and the mating gear or a control gear is mounted on the other spindle. Tooth contact is evaluated by coating the teeth with a gear-marking compound and running the set under light load for a short time. At the same time, the smoothness of operation is observed. Spacing errors and runout are evaluated by noting variations in the contact pattern on the teeth around the blank. Poor surface finish shows up as variations within the marked contact pattern. Tooth size is measured by locking one member and rotating a tooth of the mating member within the slot to determine the backlash.

The contact pattern is shifted lengthwise along the tooth to the inside and outside of the blank by displacing one member along its axis and in the offset direction. The amount of displacement is used as a measure of the set's adjustability.

It is normal practice for tooth spacing and runout to be measured with an additional operation on inspection equipment designed specifically for that purpose. AGMA publication 390.03 specifies allowable tolerances for spacing and runout based on diametral pitch and pitch diameter.

Double- and single-flank test equipment can be used to measure tooth-profile errors, tooth spacing, and runout. The test equipment has transducers on the work spindles, and the output data are in chart form. The output data not only provide a record of the quality of the gear set, but can also be related to gear noise.

Three-dimensional coordinate-measuring machines can be used to compare the actual gear-tooth geometry with theoretical data.

3-4 GEAR DESIGN CONSIDERATIONS

3-4-1 Application Requirements

Bevel and hypoid gears are suitable for transmitting power between shafts at practically any angle and speed. The load, speed, and special operating conditions must be defined as the first step in designing a gear set for a specific application.

A basic load and a suitable factor encompassing protection from intermittent overloads, desired life, and safety are determined from

1. The power rating of the prime mover, its overload potential, and the uniformity of its output torque
2. The normal output loading, peak loads and their duration, and the possibility of stalling or severe loading at infrequent intervals
3. Inertia loads arising from acceleration or deceleration

The speed or speeds at which a gear set will operate must be known to determine inertia loads, velocity factor, type of gear required, accuracy requirements, design of mountings, and the type of lubrication.

Special operating conditions include

1. Noise-level limitations
2. High ambient temperature
3. Presence of corrosive elements
4. Abnormal dust or abrasive atmosphere
5. Extreme, repetitive shock loading or reversing
6. Operating under variable alignment
7. Gearing exposed to weather
8. Other conditions that may affect the operation of the set

Notes · Drawings · Ideas

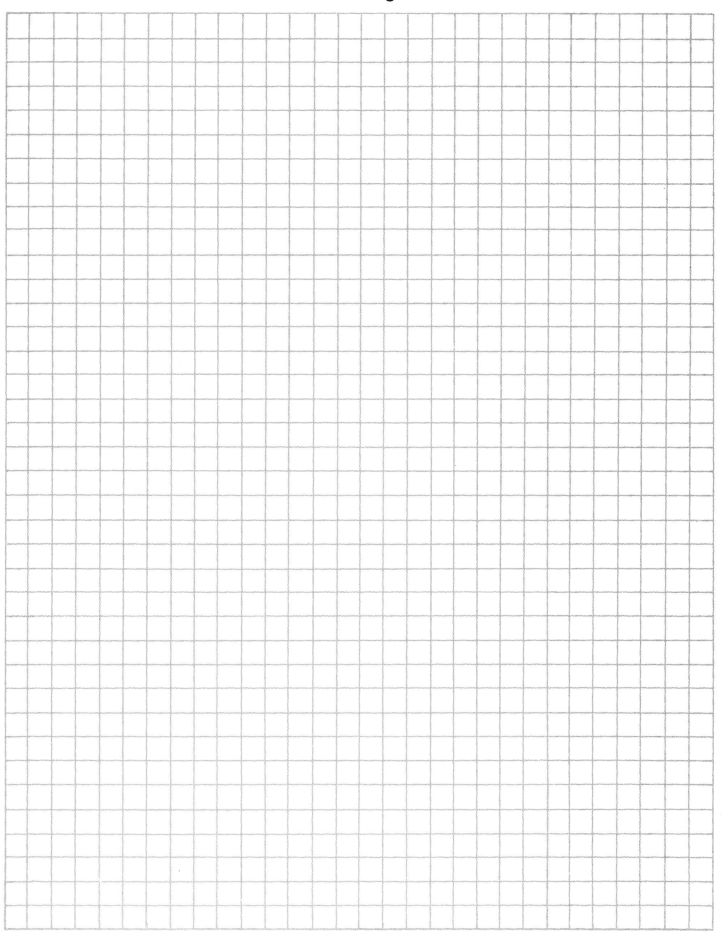

3-4-2 Selection of Type of Gear

Straight-bevel gears are recommended for peripheral speeds up to 1000 feet per minute (ft/min) where maximum smoothness and quietness are not of prime importance. However, ground straight bevels have been successfully used at speeds up to 15 000 ft/min. Plain bearings may be used for radial and axial loads and usually result in a more compact and less expensive design. Since straight-bevel gears are the simplest to calculate, set up, and develop, they are ideal for small lots.

Spiral-bevel gears are recommended where peripheral speeds are in excess of 1000 ft/min or 1000 revolutions per minute (rpm). Motion is transmitted more smoothly and quietly than with straight-bevel gears. So spiral-bevel gears are preferred also for some lower-speed applications. Spiral bevels have greater load sharing, resulting from more than one tooth in contact.

Zerol bevel gears have little axial thrust as compared to spiral-bevel gears and can be used in place of straight-bevel gears. The same qualities as defined under straight bevels apply to Zerol bevels. Because Zerol bevel gears are manufactured on the same equipment as spiral-bevel gears, Zerol bevel gears are preferred by some manufacturers. They are more easily ground because of the availability of bevel grinding equipment.

Hypoid gears are recommended where peripheral speeds are in excess of 1000 ft/min and the ultimate in smoothness and quietness is required. They are somewhat stronger than spiral bevels. Hypoids have lengthwise sliding action which enhances the lapping operation but makes them slightly less efficient than spiral-bevel gears.

3-4-3 Estimated Gear Size

Figures 3-10 and 3-11 relate size of bevel and hypoid gears to gear torque, which should be taken at a value corresponding to maximum sustained peak, or one-half peak as outlined below.

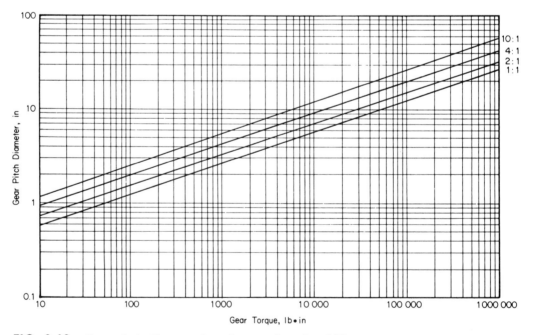

FIG. 3-10 Gear pitch diameter based on surface durability.

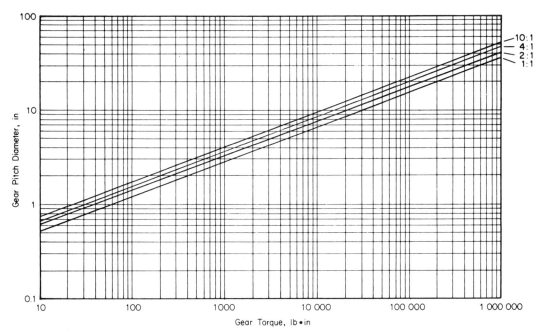

FIG. 3-11 Gear pitch diameter based on bending strength.

If the total duration of the peak load exceeds 10 000 000 cycles during the expected life of the gear, use the value of this peak load for estimating gear size. If, however, the total duration of the peak load is less than 10 000 000 cycles, use one-half the peak load or the value of the highest sustained load, whichever is greater.

Given gear torque and the desired gear ratio, the charts give gear pitch diameter. The charts are based on case-hardened steel and should be used as follows:

1. For other materials, multiply the gear pitch diameter by the material factor from Table 3-1.

TABLE 3-1 Material Factors C_M

Gear		Pinion		Material factor C_M
Material	Hardness	Material	Hardness	
Case-hardened steel	58 R_C†	Case-hardened steel	60 R_C†	0.85‡
Case-hardened steel	55 R_C†	Case-hardened steel	55 R_C†	1.00
Flame-hardened steel	50 R_C†	Case-hardened steel	55 R_C†	1.05
Flame-hardened steel	50 R_C†	Flame-hardened steel	50 R_C†	1.05
Oil-hardened steel	375–425 H_B	Oil-hardened steel	375–425 H_B	1.20
Heat-treated steel	250–300 H_B	Case-hardened steel	55 R_C†	1.45
Heat-treated steel	210–245 H_B	Heat-treated steel	245–280 H_B	1.65
Cast iron		Case-hardened steel	55 R_C†	1.95
Cast iron		Flame-hardened steel	50 R_C†	2.00
Cast iron		Annealed steel	160–200 H_B	2.10
Cast iron		Cast iron		3.10

†Minimum values.
‡Gears must be file-hard.

2. For general industrial gearing, the preliminary gear size is based on surface durability.
3. For straight-bevel gears, multiply the gear pitch diameter by 1.2; for Zerol bevel gears, multiply the gear pitch diameter by 1.3.
4. For high-capacity spiral-bevel and hypoid gears, the preliminary gear size is based on both surface capacity and bending strength. Choose the larger of the gear diameters, based on the durability chart and the strength chart.
5. For high-capacity ground spiral-bevel and hypoid gears, the gear diameter from the durability chart should be multiplied by 0.80.
6. For hypoid gears, multiply the gear pitch diameter by $D/(D + E)$.
7. Statically loaded gears should be designed for bending strength rather than surface durability. For statically loaded gears subject to vibration, multiply the gear diameter from the strength chart by 0.70. For statically loaded gears not subject to vibration, multiply the gear diameter from the strength chart by 0.60.
8. Estimated pinion diameter is $d = Dn/N$.

3-4-4 Number of Teeth

Figure 3-12 gives the recommended tooth numbers for spiral-bevel and hypoid gears. Figure 3-13 gives the recommended tooth numbers for straight-bevel and Zerol bevel gears. However, within limits, the selection of tooth numbers can be made in an arbitrary manner.

More uniform gears can be obtained in the lapping process if a common factor between gear and pinion teeth is avoided. Automotive gears are generally designed with fewer pinion teeth. Table 3-2 indicates recommended tooth numbers for automotive spiral-bevel and hypoid drives.

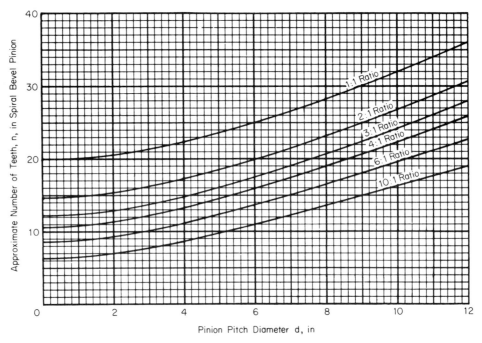

FIG. 3-12 Recommended tooth numbers for spiral-bevel and hypoid gears.

Notes · Drawings · Ideas

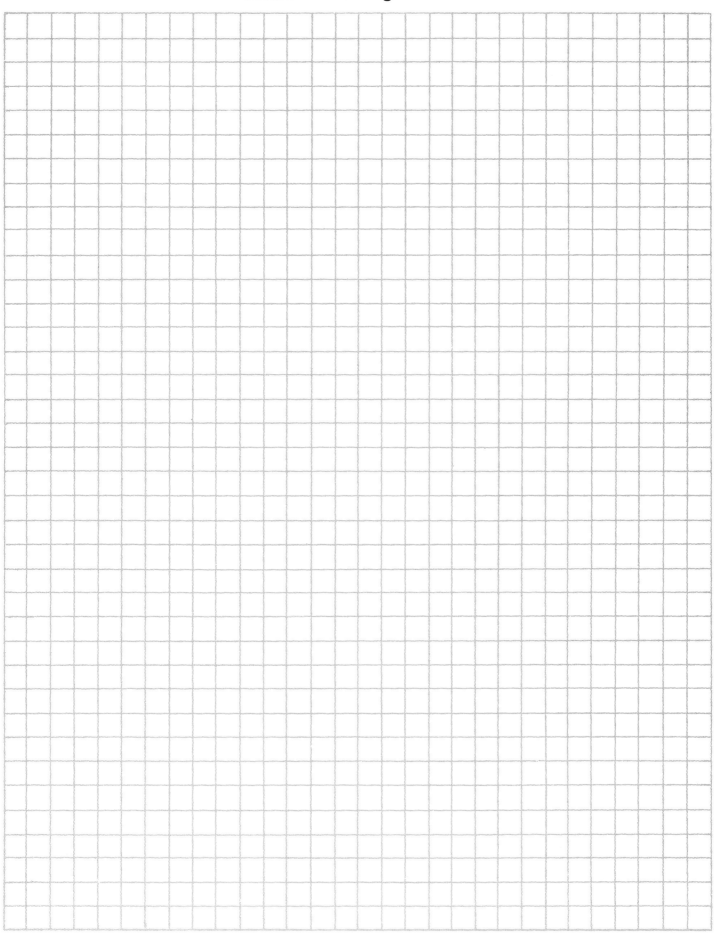

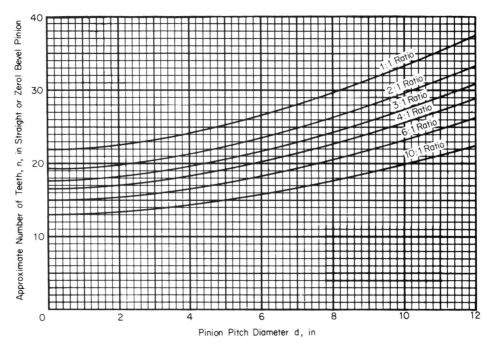

FIG. 3-13 Recommended tooth numbers for straight- and Zerol bevel gears.

3-4-5 Face Width

The face width should not exceed 30 percent of the cone distance for straight-bevel, spiral-bevel, and hypoid gears and should not exceed 25 percent of the cone distance for Zerol bevel gears. In addition, it is recommended that the face width F be limited to

$$F \leq \frac{10}{P_d}$$

TABLE 3-2 Recommended Tooth Numbers for Automotive Applications

Approximate ratio	Preferred no. pinion teeth	Allowable range
1.50/1.75	14	12 to 16
1.75/2.00	13	11 to 15
2.0/2.5	11	10 to 13
2.5/3.0	10	9 to 11
3.0/3.5	10	9 to 11
3.5/4.0	10	9 to 11
4.0/4.5	9	8 to 10
4.5/5.0	8	7 to 9
5.0/6.0	7	6 to 8
6.0/7.5	6	5 to 7
7.5/10.0	5	5 to 6

The design chart in Fig. 3-14 gives the approximate face width for straight-bevel, spiral-bevel, and hypoid gears. For Zerol bevel gears, the fact width given by this chart should be multiplied by 0.83.

3-4-6 Diametral Pitch

The diametral pitch is now calculated by dividing the number of teeth in the gear by the gear pitch diameter. Because tooling for bevel gears is not standardized according to pitch, it is not necessary that the diametral pitch be an integer.

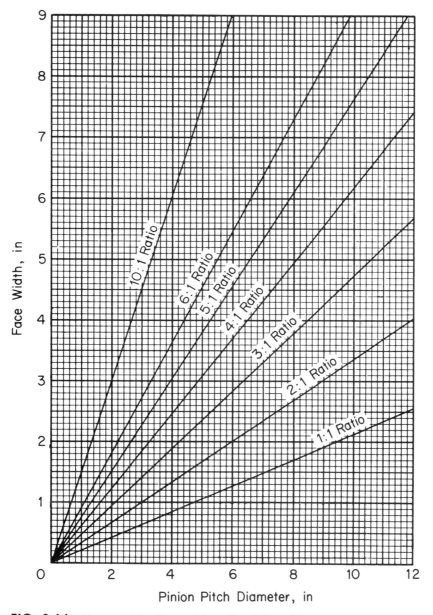

FIG. 3-14 Face width of spiral-bevel and hypoid gears.

3-4-7 Hypoid Offset

In the design of hypoid gears, the offset is designated as being above or below center. Figure 3-15a and b illustrates the below-center position, and Fig. 3-15c and d illustrates the above-center position. In general, the shaft offset for power drives should not exceed 25 percent of the gear pitch diameter, and on very heavy loaded gears, the offsct should be limited to 12.5 percent of the gear pitch diameter.

Hypoid pinions are larger in diameter than the corresponding spiral-bevel pinion. This increase in diameter may be as great as 30 percent, depending on the offset, spiral angle, and gear ratio.

3-4-8 Spiral Angle

In designing spiral-bevel gears, the spiral angle should be sufficient to give a face-contact ratio of at least 1.25. For maximum smoothness and quietness, the face-contact ratio should be between 1.50 and 2.00. High-speed applications should be designed with a face-contact ratio of 2.00 or higher for best results. Figure 3-16 may be used to assist in the selection of the spiral angle.

For hypoid gears, the desired pinion spiral angle can be calculated by

$$\psi_P = 25 + 5 \sqrt{\frac{N}{n}} + 90 \frac{E}{D}$$

where ψ_P is in degrees.

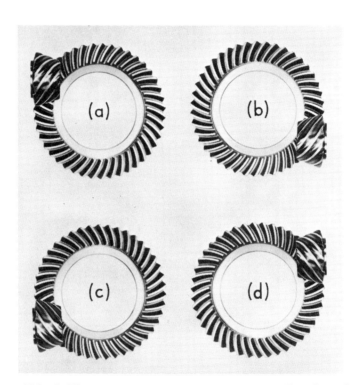

FIG. 3-15 Hypoid offset. To determine the direction of offset, always look at the gear with the pinion at the right. Thus the gear sets of (a) and (b) are both offset *below* center; similar reasoning shows that (c) and (d) are offset *above* center. (*Gleason Machine Division.*)

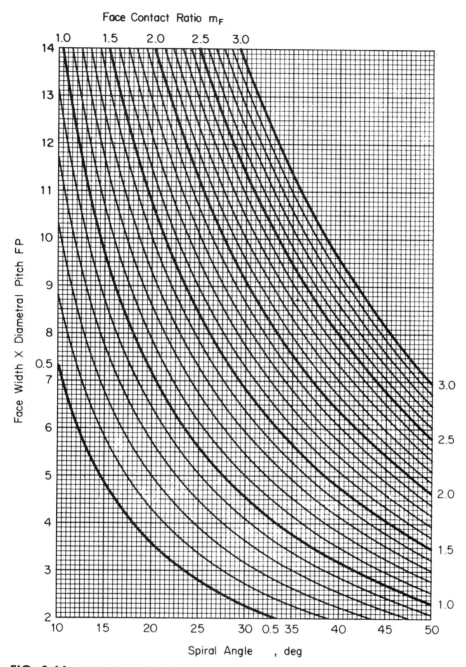

FIG. 3-16 Selection of spiral angle.

3-4-9 Pressure Angle

The commonly used pressure angle for bevel gears is 20°, although pressure angles of 22.5° and 25° are used for heavy-duty drives.

In the case of hypoids, the pressure angle is unbalanced on opposite sides of the gear teeth in order to produce equal contact ratios on the two sides. For this reason, the average pressure angle is specified for hypoids. For automotive drives use 18° or 20°, and for heavy-duty drives use 22.5° or 25°.

3-4-10 Cutter Diameter

A cutter diameter must be selected for spiral-bevel, Zerol bevel, and hypoid gears. The usual practice is to use a cutter diameter approximately equal to the gear diameter. To increase adjustability of the gear set and obtain maximum strength, a smaller cutter should be used. Cutter diameters are standardized. Therefore, Table 3-3 is included to aid in cutter selection.

3-4-11 Materials and Heat Treatment

Through-hardening steels are used when medium wear resistance and medium load-carrying capacity are desired. The following steels are some of those used and listed, beginning with the steel of lowest hardenability: AISI 1045, 1144, 4640, 4150, and 4340. When greater hardenability is required for larger gears, it is sometimes necessary to increase the carbon content of these steels or to select a different steel.

Carburized gears are used when high wear resistance and high load-carrying capacity are required. Carburizing steels used in gears normally have a carbon content of 0.10 to 0.25 percent and should have sufficient alloy content to allow hardening in the section sizes in which they are used. For low-heat-treat distortion, 4620 or 8620 might be used; if high-core hardness is desired, 9310 might be used. The following steels have been commonly used and are listed, beginning with the steel of lowest core hardenability: AISI 4620, 8620, 9310, and 4820.

Carburized gears should be specified as follows:

1. Total depth of carburized case after finishing operations
2. Surface hardness
3. Core hardness
4. Maximum case carbon content (optional)

TABLE 3-3 Standard Cutter Radii Corresponding to Various Gear Pitch Diameters for Spiral-Bevel, Zerol Bevel, and Hypoid Gears

Pitch diameter D of gear, in	Cutter radius r_c, in (mm)
3.000–5.250	1.750
3.875–6.750	2.250
4.250–7.500	2.500
5.125–9.000	3.000
6.500–11.250	3.750
7.750–13.500	4.500
9.000–15.750	5.250
10.250–18.000	6.000
12.000–21.000	7.000
13.750–24.000	8.000
15.500–27.000	9.000
18.000–31.500	10.500
21.750–60.000	(320)
27.250–75.000	(400)
34.250–100.000	(500)

Gears should be quenched from a temperature which will ensure a minimum amount of retained austenite.

Nitriding steels are used in applications which require high wear resistance with minimum distortion in heat treating. The commonly used steels are AISI 4140, 4150, and 4340. If extreme hardness and wear resistance are required, the nitralloy steels can be used. To achieve the desired results in the nitriding operation, all material should be hardened and tempered above the nitriding temperature prior to finish machining. Sharp corners should be avoided on external surfaces.

Nitrided gears should be specified as follows:

1. Total depth of nitrided case after finishing operations
2. Surface hardness
3. Core hardness

Cast iron is used in place of non-heat-treated steel where good wear resistance plus excellent machineability is required. Complicated shapes can be cast more easily from iron than they can be produced by machining from bars or forgings.

3-5 GEAR-TOOTH DIMENSIONS

3-5-1 Calculation of Basic Bevel-Gear-Tooth Dimensions

All bevel-gear-tooth dimensions are calculated in a similar manner. Therefore, straight-bevel, spiral-bevel, and Zerol bevel gears are considered as a group. In Sec. 3-4 we selected

1. Number of pinion teeth n
2. Number of gear teeth N
3. Diametral pitch P_d
4. Shaft angle Σ
5. Face width F
6. Pressure angle ϕ
7. Spiral angle ψ
8. Hand of spiral (pinion), left-hand/right-hand (LH/RH)
9. Cutter radius r_c

The formulas in Table 3-4 are now used to calculate the blank and tooth dimensions.

3-5-2 Tooth Taper

Spiral-bevel- and hypoid-gear blanks are designed by one of four methods—standard taper, duplex taper, tilted root line, or uniform depth.

Standard taper is the case where the root lines of mating members, if extended, would intersect the pitch-cone apex. The tooth depth changes in proportion to the cone distance.

Duplex taper is the case where the root lines are tilted so that the slot width is

TABLE 3-4 Formulas for Computing Blank and Tooth Dimensions

Item	Item no.	Member	Formula
Pitch diameter	1	Pinion	$d = \dfrac{n}{P_d}$
		Gear	$D = \dfrac{N}{P_d}$
Pitch angle	2	Pinion	$\gamma = \tan^{-1} \dfrac{\sin \Sigma}{N/n + \cos \Sigma}$
		Gear	$\Gamma = \Sigma - \gamma$
Outer cone distance	3	Both	$A_o = \dfrac{0.50 D}{\sin \Gamma}$
Mean cone distance	4	Both	$A_m = A_o - 0.5 F$
Depth factor k_1	5	Both	Table 3-5
Mean working depth	6	Both	$h = \dfrac{k_1 A_m}{P_d A_o} \cos \psi$
Clearance factor k_2	7	Both	Table 3-6
Clearance	8	Both	$c = k_2 h$
Mean whole depth	9	Both	$h_m = h + c$
Equivalent 90° ratio	10	Both	$m_{90} = \sqrt{\dfrac{N \cos \gamma}{n \cos \Gamma}}$
Mean addendum factor C_1	11	Both	Table 3-7
Mean circular pitch	12	Both	$P_m = \dfrac{\pi A_m}{P_d A_o}$
Mean addendum	13	Pinion	$a_P = h - a_G$
		Gear	$a_G = C_1 h$
Mean dedendum	14	Pinion	$b_P = h_m - a_P$
		Gear	$b_G = h_m - a_G$
Sum of dedendum angles	15	Both	$\Sigma \delta$ (see Sec. 3-5-2)

TABLE 3-4 Formulas for Computing Blank and Tooth Dimensions (*Continued*)

Item	Item no.	Member	Formula
Dedendum angle	16	Pinion Gear	δ_P (see Sec. 3-5-2) δ_G (see Sec. 3-5-2)
Face angle of blank	17	Pinion Gear	$\gamma_o = \gamma + \delta_G$ $\Gamma_o = \Gamma + \delta_P$
Root angle of blank	18	Pinion Gear	$\gamma_R = \gamma - \delta_P$ $\Gamma_R = \Gamma - \delta_G$
Outer addendum	19	Pinion Gear	$a_{oP} = a_P + 0.5F \tan \delta_G$ $a_{oG} = a_G + 0.5F \tan \delta_P$
Outer dedendum	20	Pinion Gear	$b_{oP} = b_P + 0.5F \tan \delta_P$ $b_{oG} = b_G + 0.5F \tan \delta_G$
Outer working depth	21	Both	$h_k = a_{oP} + a_{oG}$
Outer whole depth	22	Both	$h_t = a_{oP} + b_{oP}$
Outside diameter	23	Pinion Gear	$d_o = d + 2a_{oP} \cos \gamma$ $D_o = D + 2a_{oG} \cos \Gamma$
Pitch apex to crown	24	Pinion Gear	$x_o = A_o \cos \gamma - a_{oP} \sin \gamma$ $X_o = A_o \cos \Gamma - a_{oG} \sin \Gamma$
Mean diametral pitch	25	Both	$P_{dm} = P_d \dfrac{A_o}{A_m}$
Mean pitch diameter	26	Pinion Gear	$d_m = \dfrac{n}{P_{dm}}$ $D_m = \dfrac{N}{P_{dm}}$
Thickness factor K	27	Both	Fig. 3-17
Mean normal circular thickness	28	Pinion Gear	$t_n = P_m \cos \psi - T_n$ $T_n = \dfrac{P_m}{2 \cos \psi} - (a_P - a_G) \tan \phi + \dfrac{K \cos \psi}{P_{dm} \tan \phi}$

TABLE 3-4 Formulas for Computing Blank and Tooth Dimensions (*Continued*)

Item	Item no.	Member	Formula
Outer normal backlash allowance	29	Both	B (Table 3-8)
Mean normal chordal thickness	30	Pinion	$t_{nc} = t_n - \dfrac{t_n^3}{6d_m^2} - 0.5B\dfrac{A_m}{A_o}\sec\phi$
		Gear	$T_{nc} = T_n - \dfrac{T_n^3}{6D_m^2} - 0.5B\left(\dfrac{A_m}{A_o}\right)\sec\phi$
Mean chordal addendum	31	Pinion	$a_{cP} = a_P + \dfrac{t_n^2 \cos\gamma}{4d_m}$
		Gear	$a_{cG} = a_G + \dfrac{T_n^2 \cos\Gamma}{4D_m}$

constant. This condition permits each member of a pair to be finished in one operation using circular cutters which cut in one slot.

Tilted root-line taper is a compromise between duplex and standard taper. The blanks are designed with duplex taper except when the taper becomes excessive. When the taper is 1.3 times standard taper or greater, 1.3 times standard taper is used.

Uniform-depth taper is the case where the root lines are not tilted. The tooth depth is uniform from the inside to the outside of the blank.

TABLE 3-5 Depth Factor

Type of gear	No. pinion teeth	Depth factor k_1
Straight bevel	12 and higher	2.000
Spiral bevel	12 and higher	2.000
	11	1.995
	10	1.975
	9	1.940
	8	1.895
	7	1.835
	6	1.765
Zerol bevel	13 and higher	2.000
Hypoid	11 and higher	4.000
	10	3.900
	9	3.8
	8	3.7
	7	3.6
	6	3.5

Notes · Drawings · Ideas

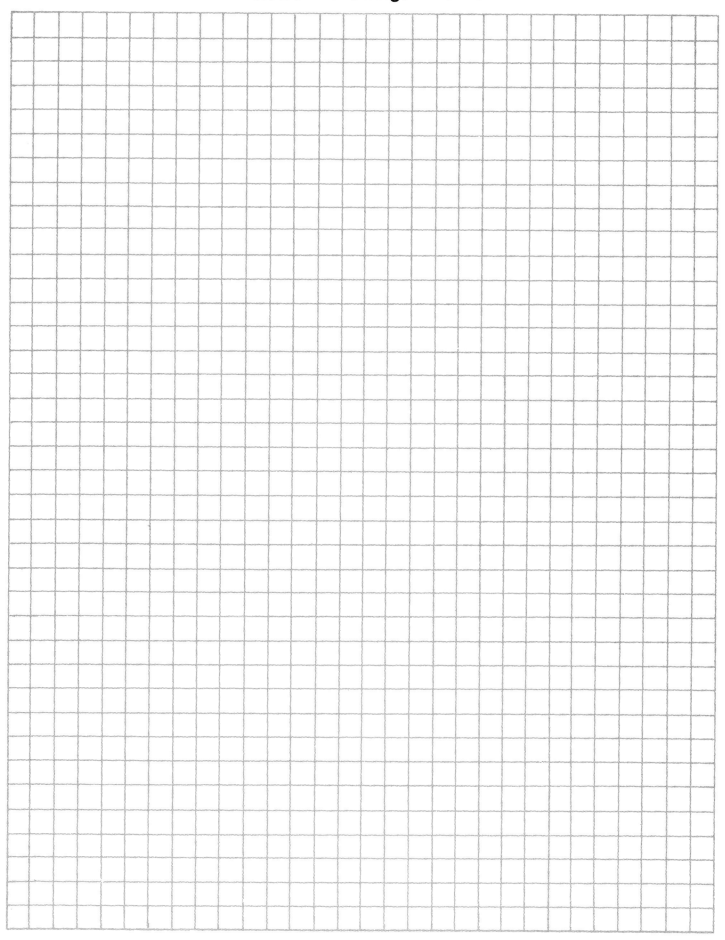

TABLE 3-6 Clearance Factors

Type of gear	Clearance factor k_2
Straight bevel	0.140
Spiral bevel	0.125
Zerol bevel	0.110
Hypoid	0.150

In many cases, the type of taper depends on the manufacturing method. Before selecting a tooth taper, you should consult with the manufacturer to ensure compatibility between the design and the cutting method.

Straight-bevel gears are usually designed with standard taper. Zerol bevel gears are usually designed with duplex taper.

The formulas used to calculate the sum of dedendum angles and the dedendum angles are shown in Table 3-9.

3-5-3 Hypoid Dimensions

The geometry of hypoid gears is complicated by the offset between the axes of the mating members. Therefore a separate set of calculation formulas is needed.

The starting data are the same as for bevel gears with the following exceptions:

1. Hypoid offset E is required.
2. Pinion spiral angle ψ_P is specified.

The formulas in Table 3-10 are now used to calculate the blank and tooth dimensions.

TABLE 3-7 Mean Addendum Factor

Type of gear	No. pinion teeth	Mean addendum factor C_1
Straight bevel	12 and higher	C_1†
Spiral bevel	12 and higher	C_1†
	11	0.490
	10	0.435
	9	0.380
	8	0.325
	7	0.270
	6	0.215
Zerol bevel	13 and higher	C_1†
Hypoid	21 and higher	C_1†
	9 to 20	0.170
	8	0.150
	7	0.130
	6	0.110

†Use $C_1 = 0.270 + 0.230/(m_{90})^2$.

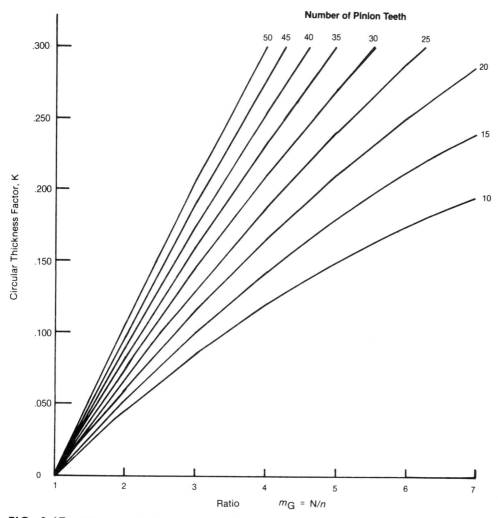

FIG. 3-17 Circular thickness factor. These curves were plotted from the equation.

$$K = -0.088 + 0.092 m_G - 0.004 m_G^2 + 0.0016(n - 30)(m_G - 1)$$

3-5-4 AGMA References†

The following AGMA standards are helpful in designing bevel and hypoid gears:

AGMA System for Spiral Bevel Gears	209.04
AGMA System for Straight Bevel Gears	208.03
AGMA System for Zerol Bevel Gears	202.03
AGMA Design Manual for Bevel Gears	330.01

---

†The notation and units used in this chapter are the same as those used in the AGMA standards. These differ in some respect to those used in other chapters of this workbook.

TABLE 3-8 Minimum Normal Backlash Allowance†

Range of diametral pitch, teeth/in	Allowance, in (for AGMA quality number range)	
	4 to 9	10 to 13
1.00–1.25	0.032	0.024
1.25–1.50	0.027	0.020
1.50–2.00	0.020	0.015
2.00–2.50	0.016	0.012
2.50–3.00	0.013	0.010
3.00–4.00	0.010	0.008
4.00–5.00	0.008	0.006
5.00–6.00	0.006	0.005
6.00–8.00	0.005	0.004
8.00–10.00	0.004	0.003
10.00–12.00	0.003	0.002
12.00–16.00	0.003	0.002
16.00–20.00	0.002	0.001
20.00–25.00	0.002	0.001

†Measured at outer cone in inches.

TABLE 3-9 Formulas for Computing Dedendum Angles and Their Sum

Type of taper	Formula
Standard	$\Sigma\delta = \tan^{-1}\dfrac{b_P}{A_m} + \tan^{-1}\dfrac{b_G}{A_m}$ $\delta_P = \tan^{-1}\dfrac{b_P}{A_m} \qquad \delta_G = \Sigma\delta - \delta_P$
Duplex	$\Sigma\delta = \dfrac{90[1 - (A_m/r_c)\sin\psi]}{(P_d A_o \tan\phi \cos\psi)}$ $\delta_P = \dfrac{a_G}{h}\Sigma\delta \qquad \delta_G = \Sigma\delta - \delta_P$
Tilted root line	Use $\Sigma\delta = \dfrac{90[1 - (A_m/r_c)\sin\psi]}{(P_d A_o \tan\phi \cos\psi)}$ or $= 1.3\tan^{-1}\dfrac{b_P}{A_m} + 1.3\tan^{-1}\dfrac{b_G}{A_m}$ whichever is smaller. $\delta_P = \dfrac{a_G}{h} \qquad \delta_G = \Sigma\delta - \delta_P$
Uniform depth	$\Sigma\delta = 0$ $\delta_P = \delta_G = 0$

TABLE 3-10 Formulas for Computing Blank and Tooth Dimensions of Hypoid Gears

Item	No.	Formula
Pitch diameter of gear	1	$D = \dfrac{N}{P_d}$
	2	$m = \dfrac{n}{N}$
	3	$\psi_{Po} = \psi_P$
	4	$\Delta\Sigma = 90 - \Sigma$
	5	$\tan \Gamma_i = \dfrac{\cos \Delta\Sigma}{1.2(m - \sin \Delta\Sigma)}$
	6	$R = 0.5(D - F \sin \Gamma_i)$
	7	$\sin \varepsilon'_i = \dfrac{E}{R} \sin \Gamma_i$
	8	$K_1 = \tan \psi_{Po} \sin \varepsilon'_i + \cos \varepsilon'_i$
	9	$R_{P2} = mRK_1$
	10	$\tan \eta = \dfrac{E}{R(\tan \Gamma_i \cos \Delta\Sigma - \sin \Delta\Sigma) + R_{P2}}$ first trial
	11	$\sin \varepsilon_2 = \dfrac{E - R_{P2} \sin \eta}{R}$
	12	$\tan \gamma_2 = \dfrac{\sin \eta}{\tan \varepsilon_2 \cos \Delta\Sigma} + \tan \Delta\Sigma \cos \eta$
	13	$\sin \varepsilon'_2 = \dfrac{\sin \varepsilon_2 \cos \Delta\Sigma}{\cos \gamma_2}$
	14	$\tan \psi_{P2} = \dfrac{K_1 - \cos \varepsilon'_2}{\sin \varepsilon'_2}$
	15	$\Delta K = \sin \varepsilon'_2 (\tan \psi_{Po} - \tan \psi_{P2})$
	16	$\dfrac{\Delta R_P}{R} = m(\Delta K)$
	17	$\sin \varepsilon_1 = \sin \varepsilon_2 - \dfrac{\Delta R_P}{R} \sin \eta$
Pinion pitch angle	18	$\tan \gamma = \dfrac{\sin \eta}{\tan \varepsilon_1 \cos \Delta\Sigma} + \tan \Delta\Sigma \cos \eta$
	19	$\sin \varepsilon'_1 = \dfrac{\sin \varepsilon_1 \cos \Delta\Sigma}{\cos \gamma_1}$
Pinion spiral angle	20	$\tan \psi_P = \dfrac{K_1 + \Delta K - \cos \varepsilon'_1}{\sin \varepsilon'_1}$

TABLE 3-10 Formulas for Computing Blank and Tooth Dimensions of Hypoid Gears (*Continued*)

Item	No.	Formula
Gear spiral angle	21	$\psi_G = \psi_P - \varepsilon_1'$
Gear pitch angle	22	$\tan \Gamma = \dfrac{\sin \varepsilon_1}{\tan \eta \cos \Delta\Sigma} + \cos \varepsilon_1 \tan \Delta\Sigma$
Gear mean cone distance	23	$A_{mG} = \dfrac{R}{\sin \Gamma}$
Pinion mean cone distance	24	$\Delta R_P = R \left(\dfrac{\Delta R_P}{R} \right)$
	25	$A_{mP} = \dfrac{R_{P2} + \Delta R_P}{\sin \gamma}$
	26	$R_P = A_{mP} \sin \gamma$
Limit pressure angle	27	$-\tan \phi_{01} = \dfrac{\tan \gamma \tan \Gamma}{\cos \varepsilon_1'} \times \dfrac{A_{mP} \sin \psi_P - A_{mG} \sin \psi_G}{A_{mP} \tan \gamma + A_{mG} \tan \Gamma}$
	28	$\text{Den} = -\tan \phi_{01} \left(\dfrac{\tan \psi_P}{A_{mP} \tan \gamma} + \dfrac{\tan \psi_G}{A_{mG} \tan \Gamma} \right) + \dfrac{1}{A_{mP} \cos \psi_P} - \dfrac{1}{A_{mG} \cos \psi_G}$
	29	$r_{c1} = \dfrac{\sec \phi_{01} (\tan \psi_P - \tan \phi_G)}{\text{Den}}$
	30	$\left| \dfrac{r_c}{r_{c1}} - 1 \right| \leq 0.01$ Loop back to no. 10 and change η until satisfied.
Gear pitch apex beyond crossing point	31	$Z_P = A_{mP} \tan \gamma \sin \Gamma - \dfrac{E \tan \Delta\Sigma}{\tan \varepsilon_1}$
	32	$Z = \dfrac{R}{\tan \Gamma} - Z_P$
Gear outer cone distance	33	$A_o = \dfrac{0.5 D}{\sin \Gamma}$
	34	$\Delta F_o = A_o - A_{mG}$
Depth factor	35	k_1 (see Table 3-5)
Addendum factor	36	C_1 (see Table 3-7)
Mean working depth	37	$h = \dfrac{k_1 R \cos \psi_G}{N}$
Mean addendum	38	$a_P = h - a_G \quad a_G = C_1 h$

TABLE 3-10 Formulas for Computing Blank and Tooth Dimensions of Hypoid Gears (*Continued*)

Item	No.	Formula
Clearance factor	39	k_2 (see Table 3-6)
Mean dedendum	40	$b_P = b_G + a_G - a_P \qquad b_G = h(1 + k_2 - C_1)$
Clearance	41	$c = k_2 h$
Mean whole depth	42	$h_m = a_G + b_G$
Sum of dedendum angle	43	$\Sigma\delta$ (see Sec. 3-5-2)
Gear dedendum angle	44	δ_G (see Sec. 3-5-2)
Gear addendum angle	45	$\alpha_G = \Sigma\delta - \delta_G$
Gear outer addendum	46	$a_{oG} = a_G + \Delta F_o \sin \alpha_G$
Gear outer dedendum	47	$b_{oG} = b_G + \Delta F_o \sin \delta_G$
Gear whole depth	48	$h_t = a_{oG} + b_{oG}$
Gear working depth	49	$h_k = h_{tG} - c$
Gear root angle	50	$\Gamma_R = \Gamma - \delta_G$
Gear face angle	51	$\Gamma_o = \Gamma + \alpha_G$
Gear outside diameter	52	$D_o = 2a_{oG} \cos \Gamma + D_G$
Gear crown to crossing point	53	$X_o = Z_P + \Delta F_o \cos \Gamma - a_{oG} \sin \Gamma$
Gear root apex beyond crossing point	54	$Z_R = Z + \dfrac{A_{mG} \sin \delta_G - b_G}{\sin \Gamma_R}$
Gear face apex beyond crossing point	55	$Z_o = Z + \dfrac{A_{mG} \sin \alpha_G - a_G}{\sin \Gamma_o}$
	56	$Q_R = \dfrac{A_{mG} \cos \delta_G}{\cos \Gamma_R} - Z$
	57	$Q_o = \dfrac{A_{mG} \cos \alpha_G}{\cos \Gamma_o} - Z$
	58	$\tan \xi_R = \dfrac{E \tan \Delta\Sigma}{Q_R}$

TABLE 3-10 Formulas for Computing Blank and Tooth Dimensions of Hypoid Gears (*Continued*)

Item	No.	Formula
Gear face apex beyond crossing point (*continued*)	59	$\tan \xi_o = \dfrac{E \tan \Delta\Sigma}{Q_o}$
	60	$\sin(\varepsilon_R + \xi_R) = \dfrac{E \cos \xi_R \tan \Gamma_R}{Q_R}$
	61	$\sin(\varepsilon_o + \xi_o) = \dfrac{E \cos \xi_o \tan \Gamma_o}{Q_o}$
Pinion face angle	62	$\sin \gamma_o = \sin \Delta\Sigma \sin \Gamma_R + \cos \Delta\Sigma \cos \Gamma_R \cos \varepsilon_R$
Pinion root angle	63	$\sin \gamma_R = \sin \Delta\Sigma \sin \Gamma_o + \cos \Delta\Sigma \cos \Gamma_o \cos \varepsilon_o$
Pinion face apex beyond crossing point	64	$G_o = \dfrac{E \sin \varepsilon_R \cos \Gamma_R - Z_R \sin \Gamma_R - c}{\sin \gamma_o}$
Pinion root apex beyond crossing point	65	$G_R = \dfrac{E \sin \varepsilon_o \cos \Gamma_o - Z_o \sin \Gamma_o - c}{\sin \gamma_R}$
	66	$\tan \lambda' = \dfrac{m \sin \varepsilon_i' \cos \Gamma}{\cos \gamma + m \cos \Gamma \cos \varepsilon_i'}$
Pinion addendum angle	67	$\alpha_P = \gamma_o - \gamma$
Pinion dedendum angle	68	$\delta_P = \gamma - \gamma_R$
Pinion whole depth	69	$h_{tP} = \dfrac{(x_o + G_o) \sin \delta_P}{\cos \gamma_o} - \sin \gamma_R (G_R - G_o)$
	70	$\Delta F_i = F - \Delta F_o$
	71	$\Delta F_{oP} = h \sin \varepsilon_R (1 - m)$
	72	$F_{oP} = \dfrac{\Delta F_o \cos \lambda'}{\cos(\varepsilon_i' - \lambda')}$
	73	$F_{iP} = \dfrac{\Delta F_i \cos \lambda'}{\cos(\varepsilon_i' - \lambda')}$
	74	$\Delta B_o = \dfrac{F_o \cos \gamma_o}{\cos \alpha_P} + \Delta F_{oP} - (b_G - c) \sin \gamma$
	75	$\Delta B_i = \dfrac{F \cos \gamma_o}{\cos \alpha_P} + \Delta F_{oP} - (b_G - c) \sin \gamma$
Pinion crown to crossing point	76	$x_o = \dfrac{E}{\tan \varepsilon_1 \cos \Delta\Sigma} - R_P \tan \gamma + \Delta B_o$

TABLE 3-10 Formulas for Computing Blank and Tooth Dimensions of Hypoid Gears (*Continued*)

Item	No.	Formula
Pinion front crown to crossing point	77	$x_i = \dfrac{E}{\tan \varepsilon_1 \cos \Delta\Sigma} - R_P \tan \gamma - \Delta B_i$
Pinion outside diameter	78	$d_o = 2 \tan \gamma_o (x_o + G_o)$
Pinion face width	79	$F_P = \dfrac{x_o - x_i}{\cos \gamma_o}$
Mean circular pitch	80	$p_m = \dfrac{\pi A_{mG}}{P_d A_o}$
Mean diametral pitch	81	$P_{dm} = P_d \dfrac{A_o}{A_{mG}}$
Thickness factor	82	K (see Fig. 3-17)
Mean pitch diameter	83	$d_m = 2A_{mP} \sin \gamma$
	84	$D_m = 2A_{mG} \sin \Gamma$
Mean normal circular thickness	85	$t_n = p_m \cos \psi_G - T_n$
	86	$T_n = 0.5 p_m \cos \psi_G - (a_P - a_G) \tan \phi + \dfrac{K \cos \psi}{P_{dm} \tan \phi}$
Outer normal backlash allowance	87	B (see Table 3-8)
Mean normal chordal thickness	88	$t_{nc} = t_n - \dfrac{t_n^3}{6d_m^2} - 0.5B \sec \phi \left(\dfrac{A_{mg}}{A_o} \right)$
	89	$T_{nc} = T_n - \dfrac{T_n^3}{6D_m^2} - 0.5B \sec \phi \left(\dfrac{A_{mG}}{A_o} \right)$
Mean chordal addendum	90	$a_{cP} = a_P + \dfrac{0.25 t_n^2 \cos \gamma}{d_m}$
	91	$a_{cG} = a_G + \dfrac{0.25 T_N^2 \cos \Gamma}{D_m}$

Notes · Drawings · Ideas

They are available through:

> American Gear Manufacturers Association
> 1500 King Street, Suite 201
> Alexandria, Virginia 22314

3-6 GEAR STRENGTH

Under ideal conditions of operation, bevel and hypoid gears have a tooth contact which utilizes the full working profile of the tooth without load concentration in any area. The recommendations and rating formulas which follow are designed for a tooth contact developed to give the correct pattern in the final mountings under full load.

3-6-1 Formulas for Contact and Bending Stress

The basic equation for contact stress in bevel and hypoid gears is

$$S_c = C_P \sqrt{\frac{2T_p C_o}{C_v} \frac{1}{FD^2} \frac{N}{n} \frac{1.2 C_m C_f}{I}} \qquad (3\text{-}1)$$

and the basic equation for bending stress is

$$S_t = \frac{2T_G K_o}{K_v} \frac{P_d}{FD} \frac{1.2 K_m}{J} \qquad (3\text{-}2)$$

where S_t = calculated tensile bending stress at root of gear tooth, pounds per square inch (lb/in²)
 S_c = calculated contact stress at point on tooth where its value will be maximum, lb/in²
 C_p = elastic coefficient of the gear-and-pinion materials combination, (lb)$^{1/2}$/in
 T_P, T_G = transmitted torques of pinion and gear, respectively, pound-inches (lb·in)
 K_o, C_o = overload factors for strength and durability, respectively
 K_v, C_v = dynamic factors for strength and durability, respectively
 K_m, C_m = load-distribution factors for strength and durability, respectively
 C_f = surface-condition factor for durability
 I = geometry factor for durability
 J = geometry factor for strength

3-6-2 Explanation of Strength Formulas and Terms

The elastic coefficient for bevel and hypoid gears with localized tooth contact pattern is given by

$$C_p = \sqrt{\frac{3}{2\pi} \frac{1}{(1-\mu_P^2)/E_P + (1-\mu_G^2)/E_G}} \qquad (3\text{-}3)$$

Notes · Drawings · Ideas

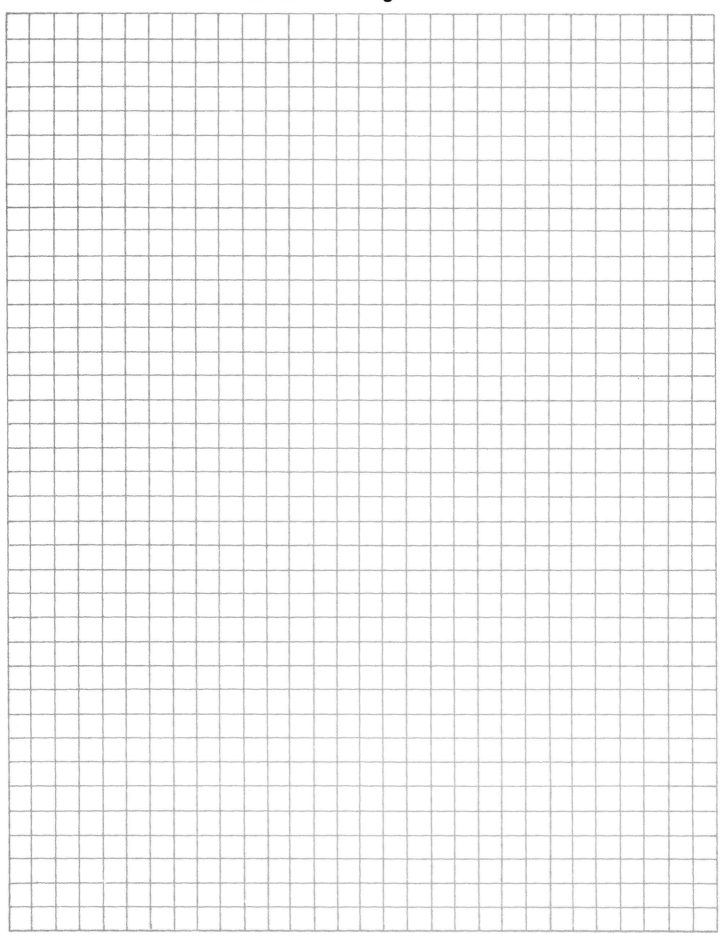

TABLE 3-11 Overload Factors K_o, C_o†

Prime mover	Character of load on driven member		
	Uniform	Medium shock	Heavy shock
Uniform	1.00	1.25	1.75
Medium shock	1.25	1.50	2.00
Heavy shock	1.50	1.75	2.25

†This table is for speed-decreasing drive; for speed-increasing drives add $0.01(N/n)^2$ to the above factors.

where μ_P, μ_G = Poisson's ratio for materials of pinion and gear, respectively (use 0.30 for ferrous materials)

E_P, E_G = Young's modulus of elasticity for materials of pinion and gear, respectively (use 30.0×10^6 lb/in² for steel)

The overload factor makes allowance for the roughness or smoothness of operation of both driving and driven units. Use Table 3-11 as a guide in selecting the overload factor.

The dynamic factor reflects the effect of inaccuracies in tooth profile, tooth spacing, and runout on instantaneous tooth loading. For gears manufactured to AGMA class 11 tolerances or higher, a value of 1.0 may be used for dynamic factor. Curve 2 in Fig. 3-18 gives the values of C_v for spiral bevels and hypoids of lower accuracy or for large, planed spiral-bevel gears. Curve 3 gives the values of C_v for bevels of lower accuracy or for large, planed straight-bevel gears.

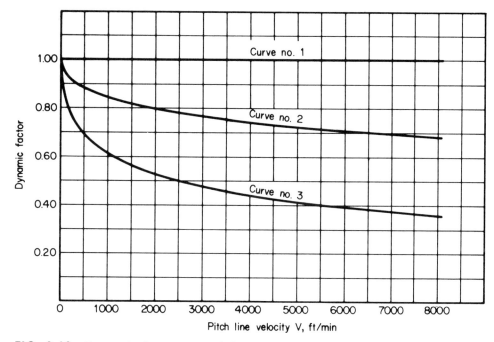

FIG. 3-18 Dynamic factors K_v and C_v.

TABLE 3-12 Load-Distribution Factors K_m, C_m

Application	Both members straddle-mounted	One member straddle-mounted	Neither member straddle-mounted
General industrial	1.00–1.10	1.10–1.25	1.25–1.40
Automotive	1.00–1.10	1.10–1.25	
Aircraft	1.00–1.25	1.10–1.40	1.25–1.50

The load-distribution factor allows for misalignment of the gear set under operating conditions. This factor is based on the magnitude of the displacements of the gear and pinion from their theoretical correct locations. Use Table 3-12 as a guide in selecting the load-distribution factor.

The surface-condition factor depends on surface finish as affected by cutting, lapping, and grinding. It also depends on surface treatment such as lubrizing. And C_f can be taken as 1.0 provided good gear manufacturing practices are followed.

The geometry factor for durability I takes into consideration the relative radius of curvature between mating tooth surfaces, load location, load sharing, effective face width, and inertia factor. The series of graphs in Figs. 3-19 to 3-36 give the dura-

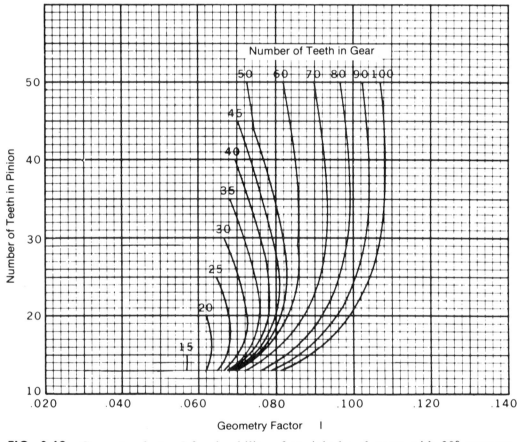

FIG. 3-19 Geometry factor I for durability of straight-bevel gears with 20° pressure angle and 90° shaft angle.

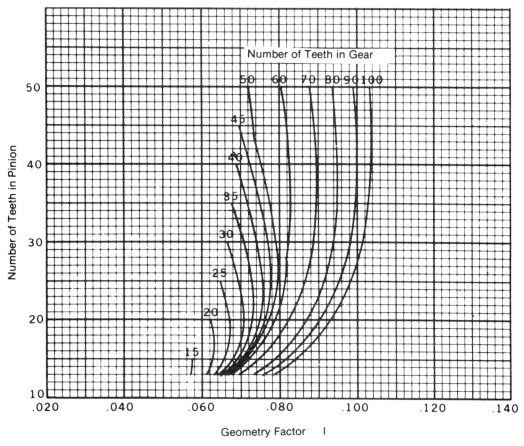

FIG. 3-20 Geometry factor I for durability of straight-bevel gears with 25° pressure angle and 90° shaft angle.

bility geometry factors for some of the most commonly used gear sets. You may have to interpolate or extrapolate between graphs to find the factor for your set.

The geometry factor for strength J takes into consideration the tooth form factor, load location, load distribution, effective face width, stress correction factor, and inertia factor. The series of graphs in Figs. 3-37 to 3-52 give the strangth geometry factors for some of the most commonly used gear sets. You may have to interpolate or extrapolate between graphs to find the factor for your set.

3-6-3 Allowable Stresses

The maximum allowable stresses are based on the properties of the material. They vary with the material, heat treatment, and surface treatment. Table 3-13 gives nominal values for allowable contact stress on gear teeth for commonly used gear materials and heat treatments. Table 3-14 gives nominal values for allowable bending stress in gear teeth for commonly used gear materials and heat treatments.

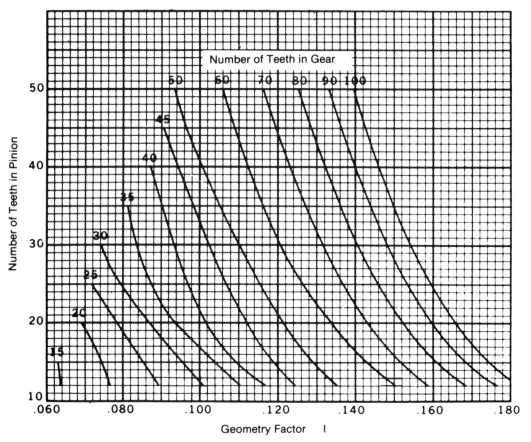

FIG. 3-21 Geometry factor I for durability of spiral-bevel gears with 20° pressure angle, 35° spiral angle, and 90° shaft angle.

Carburized case-hardened gears require a core hardness in the range of 260 to 350 H_B (26 to 37 R_C) and a total case depth in the range shown by Fig. 3-53.

The calculated contact stress S_c times a safety factor should be less than the allowable contact stress S_{ac}. The calculated bending stress S_t times a safety factor should be less than the allowable bending stress S_{at}.

3-6-4 Scoring Resistance

Scoring is a temperature-related process where the surfaces actually tend to weld together. The oil film breaks down, and the tooth surfaces roll and slide on one another, metal against metal. Friction between the surfaces causes heat which reaches the melting point of the tooth material, and scoring results. The factors which could cause scoring are the sliding velocity, surface finish, and load concentrations along with the lubricant temperature, viscosity, and application. If you follow the recommendations under Sec. 3-7-6 on lubrication and the manufacturer uses acceptable practices in processing the gears, then scoring should not be a problem.

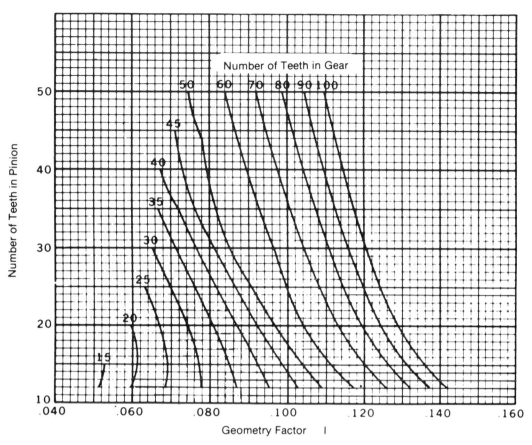

FIG. 3-22 Geometry factor I for durability of spiral-bevel gears with 20° pressure angle, 25° spiral angle, and 90° shaft angle.

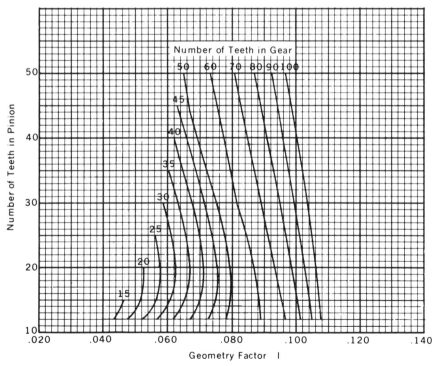

FIG. 3-23 Geometry factor I for durability of spiral-bevel gears with 20° pressure angle, 15° spiral angle, and 90° shaft angle.

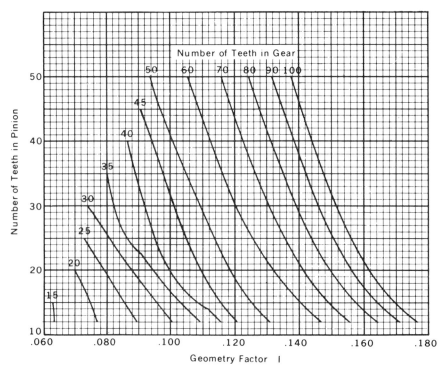

FIG. 3-24 Geometry factor I for durability of spiral-bevel gears with 25° pressure angle, 35° spiral angle, and 90° shaft angle.

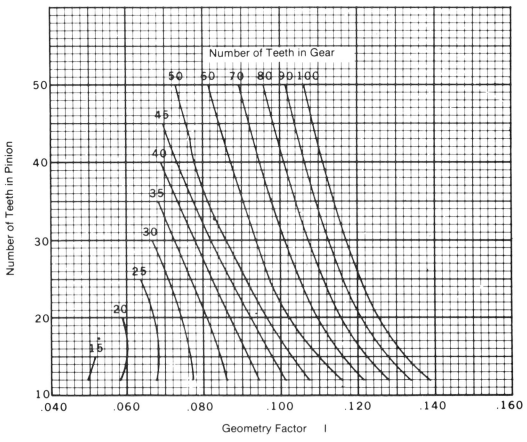

FIG. 3-25 Geometry factor I for durability of spiral-bevel gears with 25° pressure angle, 25° spiral angle, and 90° shaft angle.

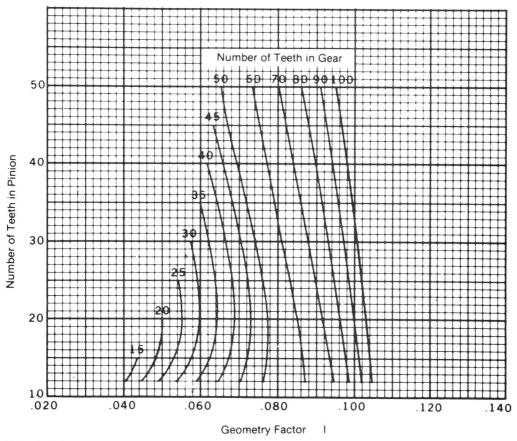

FIG. 3-26 Geometry factor I for durability of spiral-bevel gears with 25° pressure angle, 15° spiral angle, and 90° shaft angle.

3-7 DESIGN OF MOUNTINGS

The normal load on the tooth surfaces of bevel and hypoid gears may be resolved into two components: one in the direction along the axis of the gear and the other perpendicular to the axis. The direction and magnitude of the normal load depend on the ratio, pressure angle, spiral angle, hand of spiral, and direction of rotation as well as whether the gear is the driving or driven member.

3-7-1 Hand of Spiral

In general, a left-hand pinion driving clockwise (viewed from the back) tends to move axially away from the cone center; a right-hand pinion tends to move toward the center because of the oblique direction of the curved teeth. If possible, the hand of spiral should be selected so that both the pinion and the gear tend to move out of mesh, which prevents the possibility of tooth wedging because of reduced backlash. Otherwise, the hand of spiral should be selected to give an axial thrust that tends to move the pinion out of mesh. In a reversible drive, there is no choice unless the pair performs a heavier duty in one direction for a greater part of the time.

On hypoids when the pinion is below center and to the right (when you are facing

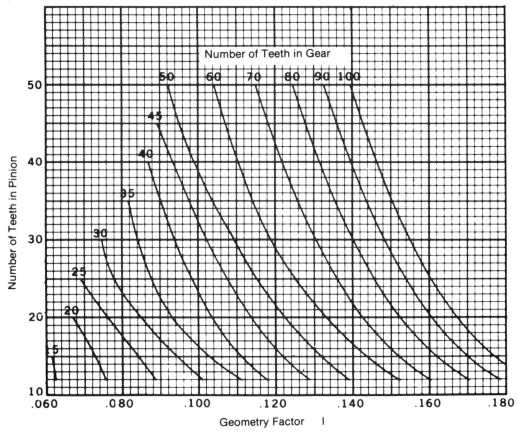

FIG. 3-27 Geometry factor I for durability of spiral-bevel gears with 16° pressure angle, 35° spiral angle, and 90° shaft angle.

the front of the gear), the pinion hand of spiral should always be left-hand. With the pinion above center and to the right, the pinion hand should always be right-hand. See Fig. 3-15.

3-7-2 Tangential Force

The tangential force on a bevel or hypoid gear is given by

$$W_{tG} = \frac{2T_G}{D_m} = \frac{126\,000P}{D_m N} \tag{3-4}$$

where T_G = gear torque, lb·in
P = power, horsepower (hp)
N = speed of gear, rpm

The tangential force on the mating pinion is given by the equation

$$W_{tP} = \frac{W_{tG} \cos \psi_P}{\cos \psi_G} = \frac{2T_P}{d_m} \tag{3-5}$$

where T_P = pinion torque in pound-inches.

Notes · Drawings · Ideas

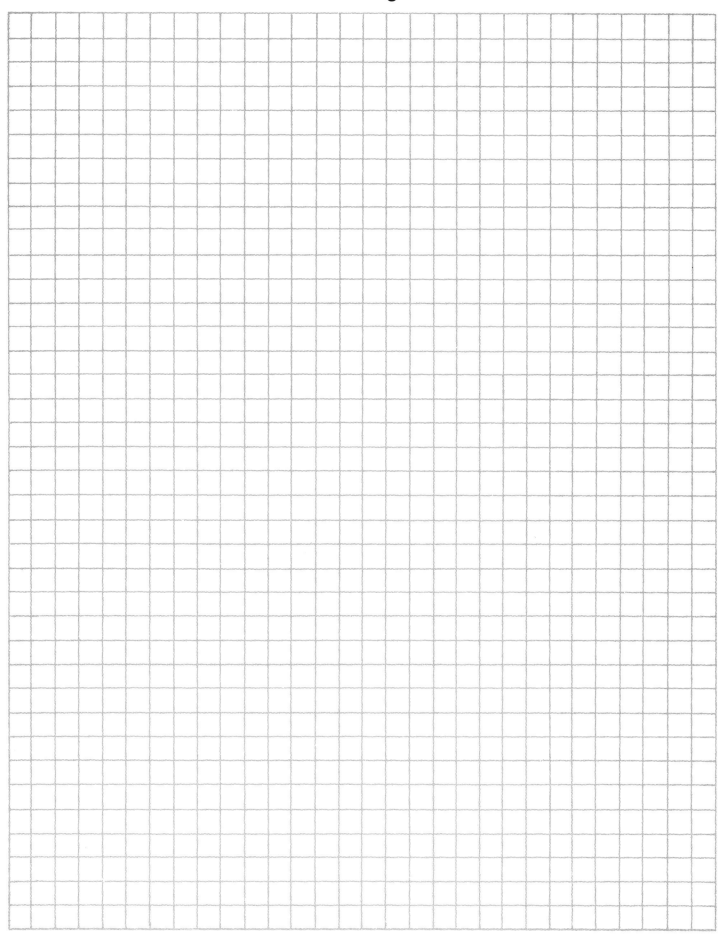

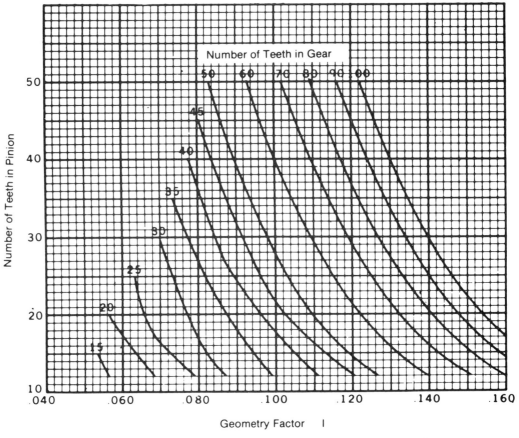

FIG. 3-28 Geometry factor I for durability of spiral-bevel gears with 20° pressure angle, 35° spiral angle, and 60° shaft angle.

3-7-3 Axial Thrust and Radial Separating Forces

The formulas that follow are used to calculate the *axial thrust force* W_x and the *radial separating force* W_R for bevel and hypoid gears. The direction of the pinion (driver) rotation should be viewed from the pinion back.

For a pinion (driver) with a *right-hand (RH) spiral with clockwise (cw) rotation*, or a *left-hand (LH) spiral with counterclockwise (ccw) rotation*, the axial and separating force components *acting on the pinion* are, respectively,

$$W_{xP} = W_{tP} \sec \psi_P (\tan \phi \sin \gamma - \sin \psi_P \cos \gamma) \tag{3-6}$$

$$W_{RP} = W_{tP} \sec \psi_P (\tan \phi \cos \gamma + \sin \psi_P \sin \gamma) \tag{3-7}$$

For a pinion (driver) with a *LH spiral and cw rotation,* or a *RH spiral with ccw rotation,* the force components *acting on the pinion* are, respectively,

$$W_{xP} = W_{tP} \sec \psi_P (\tan \phi \sin \gamma + \sin \psi_P \cos \gamma) \tag{3-8}$$

$$W_{RP} = W_{tP} \sec \psi_P (\tan \phi \cos \gamma - \sin \psi_P \sin \gamma) \tag{3-9}$$

Notes · Drawings · Ideas

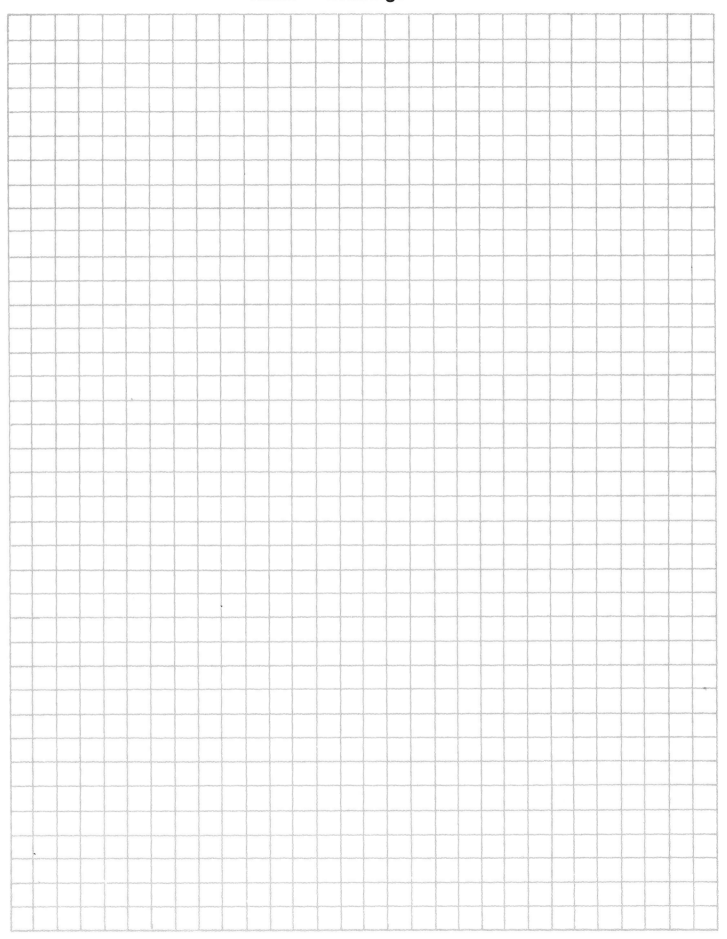

82 GEARING: A MECHANICAL DESIGNERS' WORKBOOK

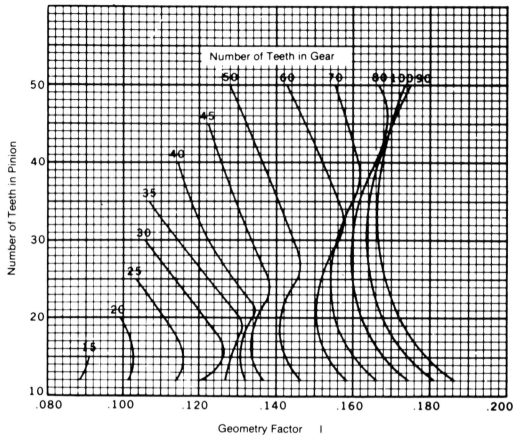

FIG. 3-29 Geometry factor I for durability of spiral-bevel gears with 20° pressure angle, 35° spiral angle, and 120° shaft angle.

For a pinion (driver) with a *RH spiral with cw rotation,* or a *LH spiral with ccw rotation,* the force components *acting on the gear* (driven) are, respectively,

$$W_{xG} = W_{tG} \sec \psi_G (\tan \phi \sin \Gamma + \sin \psi_G \cos \Gamma) \qquad (3\text{-}10)$$

$$W_{RG} = W_{tG} \sec \psi_G (\tan \phi \cos \Gamma - \sin \psi_G \sin \Gamma) \qquad (3\text{-}11)$$

For a pinion (driver) with an *LH spiral and cw rotation,* or an *RH spiral with ccw rotation,* the force components *acting on the gear* are, respectively,

$$W_{xG} = W_{tG} \sec \psi_G (\tan \phi \sin \Gamma - \sin \psi_G \cos \Gamma) \qquad (3\text{-}12)$$

$$W_{RG} = W_{tG} \sec \psi_G (\tan \phi \cos \Gamma + \sin \psi_G \sin \Gamma) \qquad (3\text{-}13)$$

These equations apply to straight-bevel, Zerol bevel, spiral-bevel, and hypoid gears. When you use them for hypoid gears, be sure that the pressure angle corresponds to the driving face of the pinion tooth.

A plus sign for Eqs. (3-6), (3-8), (3-10), and (3-12) indicates that the direction of the axial thrust is *outward,* or away from the cone center. Thus a minus sign indicates that the direction of the axial thrust is *inward,* or toward the cone center.

Notes · Drawings · Ideas

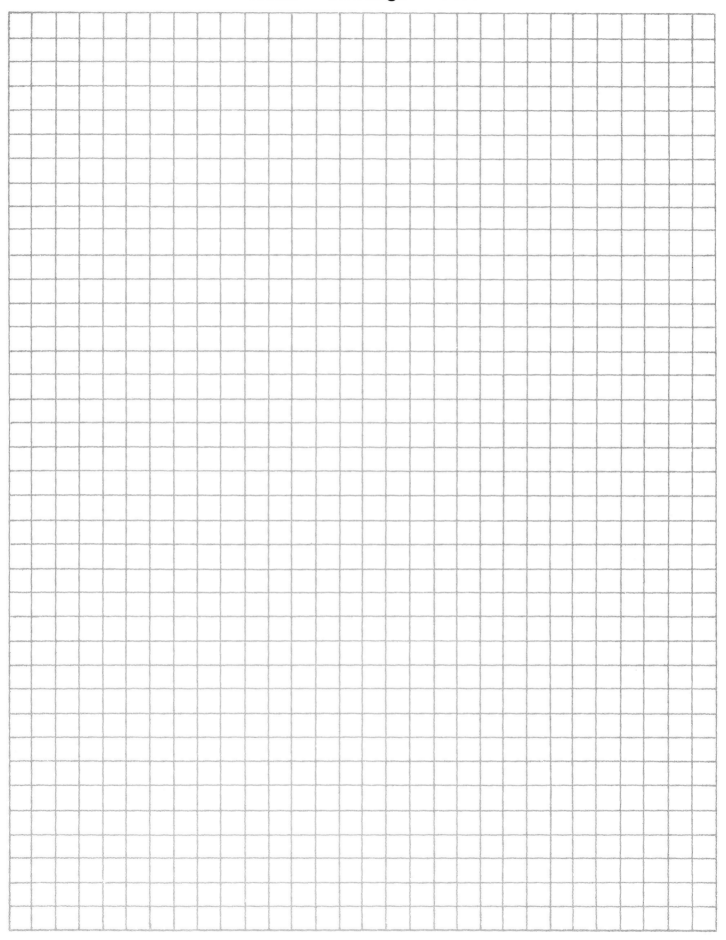

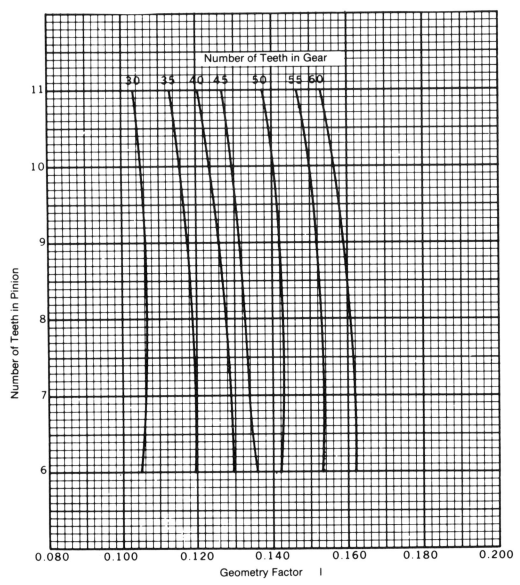

FIG. 3-30 Geometry factor *I* durability of automotive spiral-bevel gears with 20° pressure angle, 35° spiral angle, and 90° shaft angle.

A plus sign for Eqs. (3-7), (3-9), (3-11), and (3-13) indicates that the direction of the *separating* force is *away* from the mating gear. So a minus sign indicates an *attracting* force *toward* the mating member.

EXAMPLE. A hypoid-gear set consists of an 11-tooth pinion with LH spiral and ccw rotation driving a 45-tooth gear. Data for the gear are as follows: 4.286 diametral pitch, 8.965-inch (in) mean diameter, 70.03° pitch angle, 31.48° spiral angle, and 30 $\times 10^3$ lb·in torque. Pinion data are these: 1.500-in offset, 2.905-in mean diameter, concave pressure angle 18.13°, convex pressure angle 21.87°, pitch angle 19.02°, and

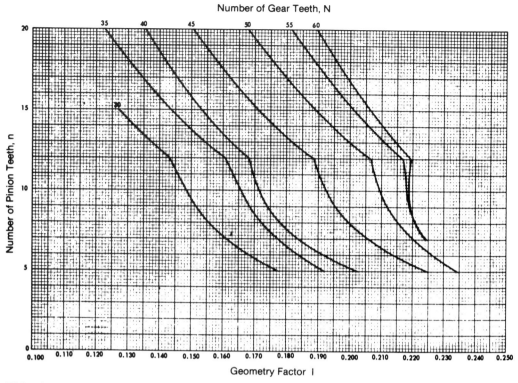

FIG. 3-31 Geometry factor I for durability of hypoid gears with 19° average pressure angle and E/D ratio of 0.10.

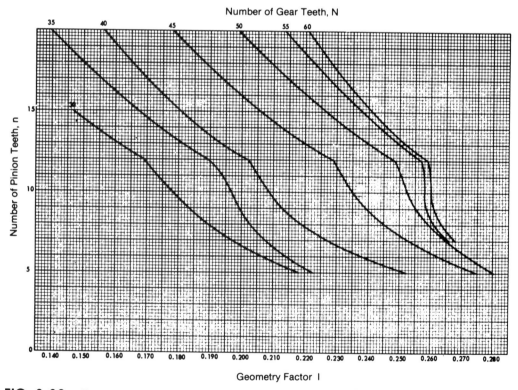

FIG. 3-32 Geometry factor I for durability of hypoid gears with 19° average pressure angle and E/D ratio of 0.15.

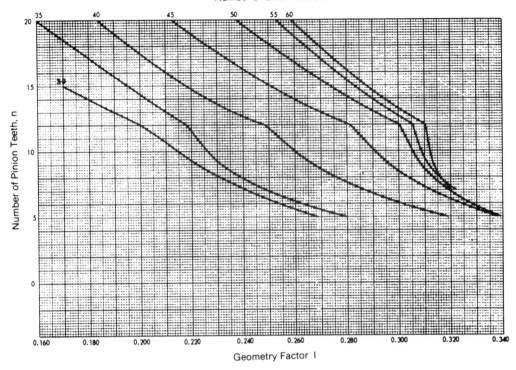

FIG. 3-33 Geometry factor I for durability of hypoid gears with 19° average pressure angle and E/D ratio of 0.20.

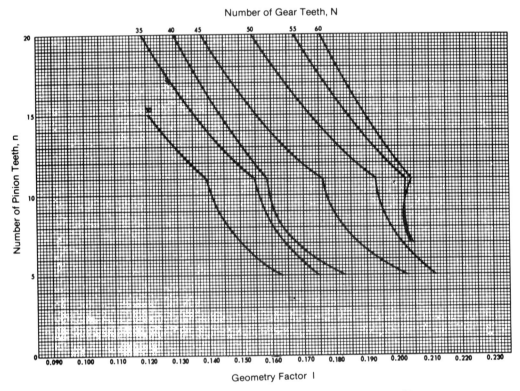

FIG. 3-34 Geometry factor I for durability of hypoid gears with $22\frac{1}{2}°$ average pressure angle and E/D ratio of 0.10.

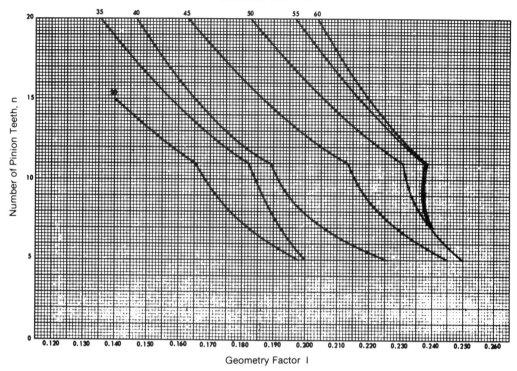

FIG. 3-35 Geometry factor I for durability of hypoid gears with $22\frac{1}{2}°$ average pressure angle and E/D ratio of 0.15.

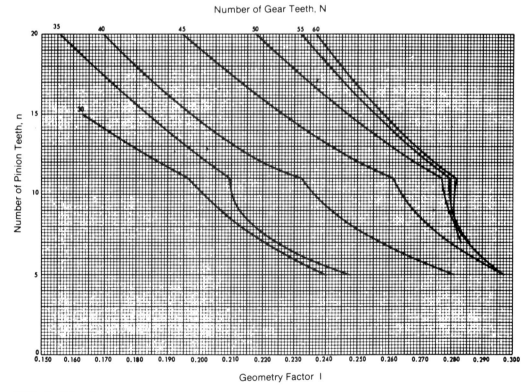

FIG. 3-36 Geometry factor I for durability of hypoid gears with $22\frac{1}{2}°$ average pressure angle and E/D ratio of 0.20.

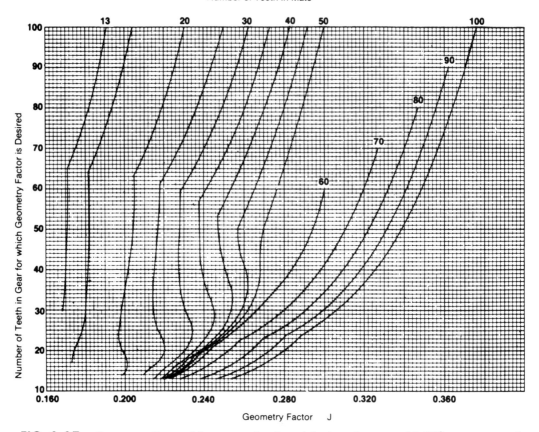

FIG. 3-37 Geometry factor J for strength of straight-bevel gears with 20° pressure angle and 90° shaft angle.

spiral angle 50°. Determine the force components and their directions for each member of the set.

Solution. From Eq. (3-4) we find the tangential load on the gear to be

$$W_{tG} = \frac{2T_G}{D_m} = \frac{2(30 \times 10^3)}{8.965} = 6693 \text{ lb}$$

Since the pinion has LH spiral angle and rotates ccw, Eqs. (3-10) and (3-11) apply for the gear. Thus

$$W_{xG} = W_{tG} \sec \psi_G (\tan \phi \sin \Gamma + \sin \psi_G \cos \Gamma)$$
$$= 6693 \sec 31.48 (\tan 18.13° \sin 70.03° + \sin 31.48° \cos 70.03°)$$
$$= 3814 \text{ lb}$$

Substituting the same values and angles into Eq. (3-11) gives $W_{RG} = -2974$ lb. Thus the thrust is outward, and the separating force is toward the mating member.

Next we find the tangential load on the pinion from Eq. (3-5):

$$W_{tP} = \frac{W_{tG} \cos \psi_P}{\cos \psi_G} = \frac{6693 \cos 50°}{\cos 31.48°} = 5045 \text{ lb}$$

Notes · Drawings · Ideas

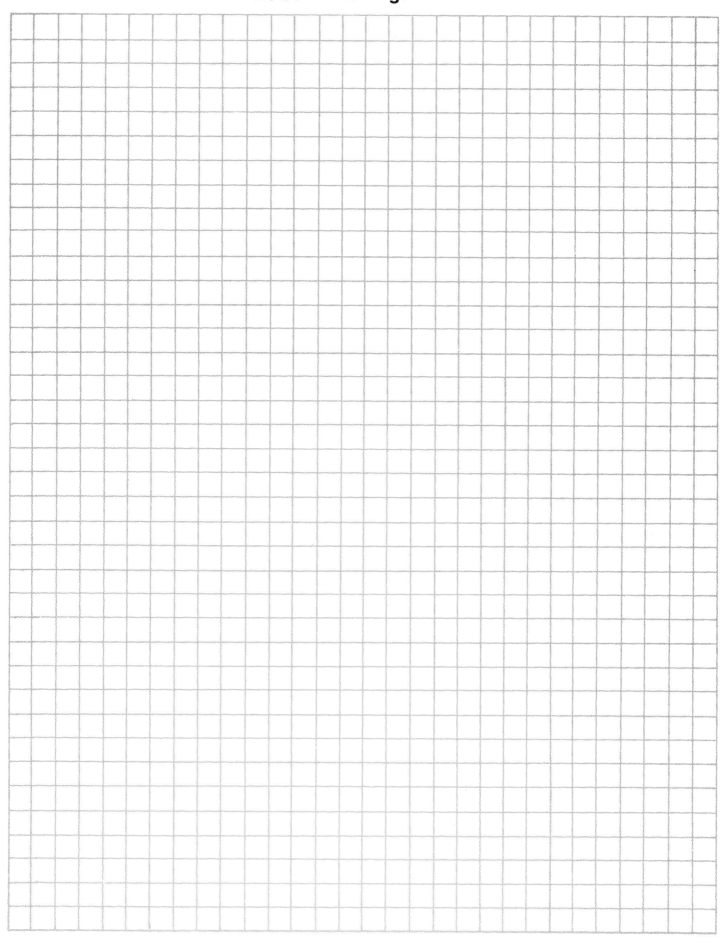

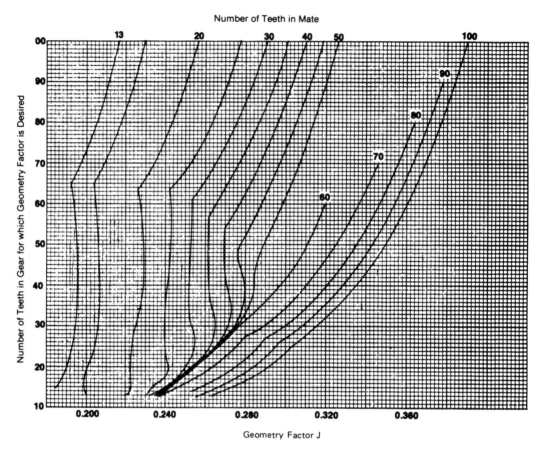

FIG. 3-38 Geometry for *J* factor strength of straight-bevel gears with 25° pressure angle and 90° shaft angle.

Equations (3-6) and (3-7) apply to the pinion:

$$W_{xP} = W_{tP} \sec \psi_P (\tan \phi \sin \gamma - \sin \psi_P \cos \gamma)$$
$$= 5045 \sec 50°(\tan 18.13° \sin 19.02° - \sin 50° \cos 19.02°)$$
$$= -4846 \text{ lb}$$

In a similar manner Eq. (3-7) gives $W_{RP} = 4389$ lb. Thus the axial thrust is inward, and the separating force away from the gear.

3-7-4 Bearing Loads

The bearings selected must be adequate to support the axial forces W_x for both directions of rotation and for the load conditions on both sides of the teeth.

Radial forces are transmitted indirectly through moment arms to the bearings. The radial bearing loads are derived from the gear separating force, the gear tangential force, and the gear thrust couple, along with the type of mounting and the bearing position.

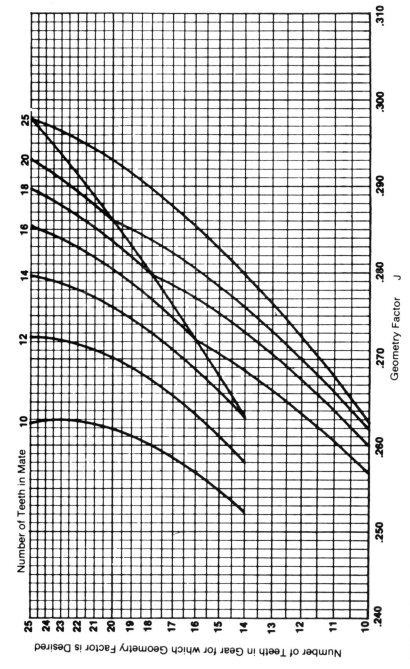

FIG. 3-39 Geometry factor J for strength of straight-bevel differential gears with $22\frac{1}{2}°$ pressure angle and 90° shaft angle.

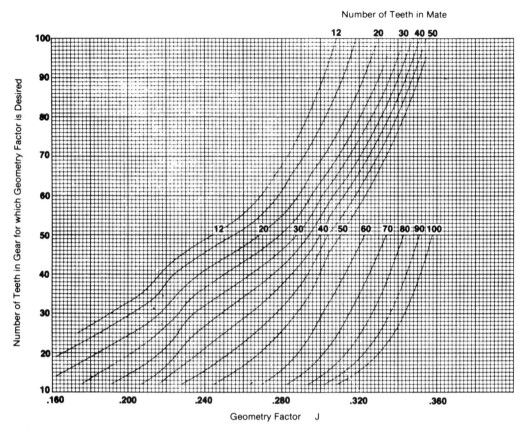

FIG. 3-40 Geometry factor J for strength of spiral-bevel gears with 20° pressure angle, 35° spiral angle, and 90° shaft angle.

3-7-5 Types of Mountings

Two types of mountings are generally used: *overhung,* where both bearings are located on the shaft behind the gear, and *straddle,* where one bearing is on either side of the gear. Because of the stiffer configuration, straddle mountings are generally used for highly loaded gears.

3-7-6 Lubrication

The lubrication system for a bevel- or hypoid-gear drive should sufficiently lubricate and adequately cool the gears and bearings. Splash lubrication is generally satisfactory for applications up to peripheral speeds of 2000 ft/min. The oil level should cover the full face of the lowest gear, and the quantity of oil should be sufficient to maintain the oil temperature within recommended limits.

Pressure lubrication is recommended for velocities above 2000 ft/min. The jets should be located to direct the stream to cover the full length of the teeth of both members, preferably close to the mesh point on the leaning side.

Experience has shown that an oil flow of 0.07 to 1.0 gallons per minute (gal/min) per 100 hp will result in an oil temperature rise of approximately 10°F.

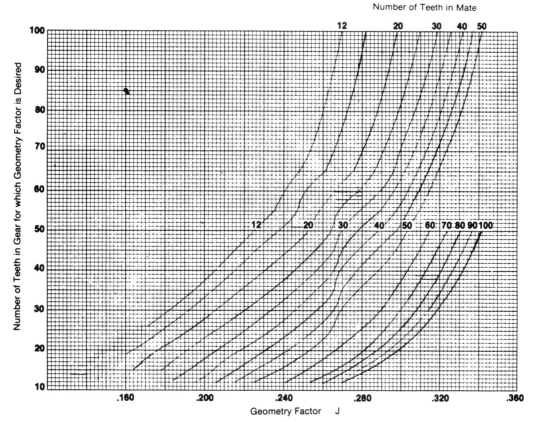

FIG. 3-41 Geometry factor J for strength of spiral-bevel gears with 20° pressure angle, 25° spiral angle, and 90° shaft angle.

Extreme-pressure (EP) lubricants are recommended for hypoid gears and for spiral-bevel gears which are subject to extreme conditions of shock, severe starting conditions, or heavy loads. The lubrication system should be fully protected against contamination by moisture or dirt. For continuous operation at temperatures above 160°F, the lubricants should be approved by the lubricant manufacturer.

In general, for a splash lubrication, an SAE 80 or 90 gear oil should be satisfactory. For a circulating system with an oil spray lubrication, SAE 20 or 30 should be satisfactory. AGMA "Specifications on Lubrication of Enclosed and Open Gearing" is a recommended guide to the type and grade of oil for various operating conditions.

3-7-7 Loaded Contact Check

With highly stressed bevel- and hypoid-gear applications such as aircraft and automotive, it is normal practice to perform a loaded contact check with the gear set assembled in its mountings. A brake load is applied to the output shafts, and the pinion member is rotated slowly at approximately 15 rpm. A marking compound is applied to the pinion and gear teeth to permit observation of the tooth contact pattern at the desired load conditions. The purpose of this test is to evaluate the rigidity of the mountings and ensure that the contact pattern remains within the tooth

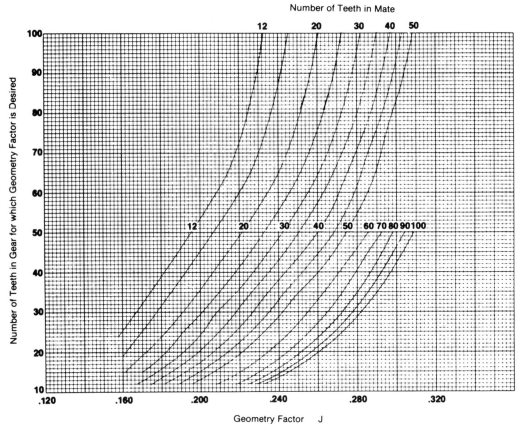

FIG. 3-42 Geometry factor *J* for strength of spiral-bevel gears with 20° pressure angle, 15° spiral angle, and 90° shaft angle.

boundaries under all load conditions. Indicators can be mounted at various positions under load. An analysis of these data can result in modifications of the mounting design or contact pattern to ensure that the contact pattern does not reach the tooth boundaries at operating loads. This will eliminate an edge contact condition which can cause noise or premature failure of the gear teeth.

3-8 COMPUTER-AIDED DESIGN

3-8-1 Computer Timesharing

A computer timesharing service is available to assist you with gear-tooth design, strength calculations, gear-tooth geometry analysis, gear manufacturing, and inspection data for bevel and hypoid gears. Contact

> Application Engineering Department
> Gleason Machine Division
> 1000 University Avenue
> Rochester, New York 14692

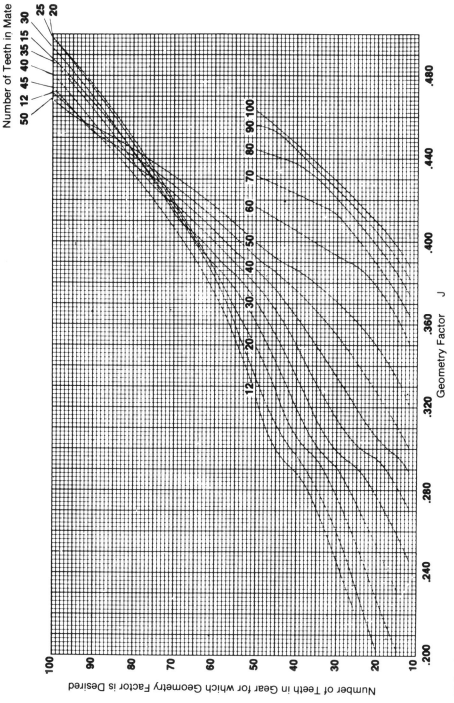

FIG. 3-43 Geometry factor J for strength of spiral-bevel gears with 25° pressure angle, 35° spiral angle, and 90° shaft angle.

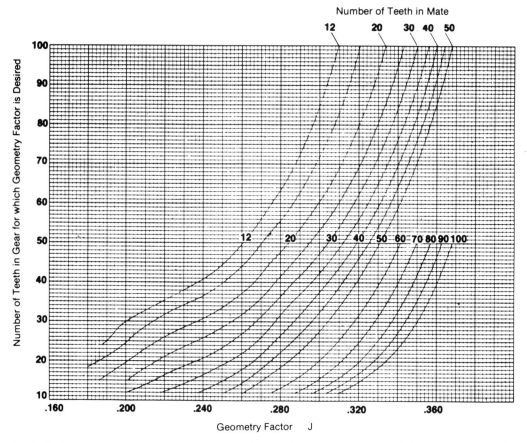

FIG. 3-44 Geometry factor J for strength of spiral-bevel gears with 20° pressure angle, 35° spiral angle, and 60° shaft angle.

3-8-2 Design Calculating Services

The Gleason Machine Division offers a calculating service which may be used as an alternative to the computer timesharing service mentioned earlier, when you require a computer analysis of the gear-tooth design.

3-8-3 Available Computer Programs

The following computer programs are available from the Gleason Machine Division to assist you with a gear-tooth design analysis:

1. *Dimension Sheet* Calculation of the basic tooth geometry, contact ratios, stress data, bearing thrust loads, and profile sliding velocities.
2. *Summary* Calculation of cutting and grinding machine setup data to produce the desired tooth geometry.
3. *Tooth Contact Analysis* A special analysis program that determines the tooth contact pattern and transmission motion errors based on specified cutting tools and gear-tooth geometry. Figure 3-54 illustrates a typical *tooth contact analysis*.

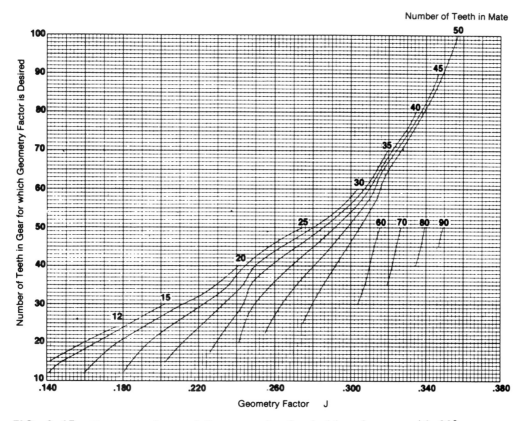

FIG. 3-45 Geometry factor J for strength of spiral-bevel gears with 20° pressure angle, 35° spiral angle, and 120° shaft angle.

4. *Undercut Check* Calculation of the location of undercut lengthwise along the tooth, along with the depth and angle of undercut relative to the tooth profile.
5. *Loaded Tooth Contact Analysis* An analysis and plot of tooth contact pattern and transmission errors as a function of gear torque. Deflections of the gear mountings may also be considered with this analysis.
6. *Finite-Element Analysis* Detailed stress data calculated based on a three-dimensional finite-element stress model which considers exact gear-tooth geometry based on cutting tool specifications, machine setup, and generating motions and mounting deflections.

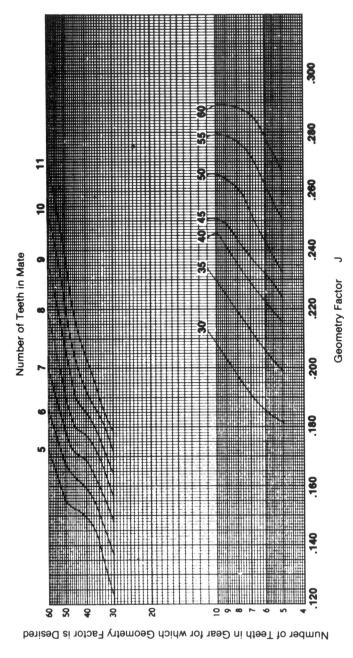

FIG. 3-46 Geometry factor J for strength of automotive spiral-bevel gears with 20° pressure angle, 35° spiral angle, and 90° shaft angle.

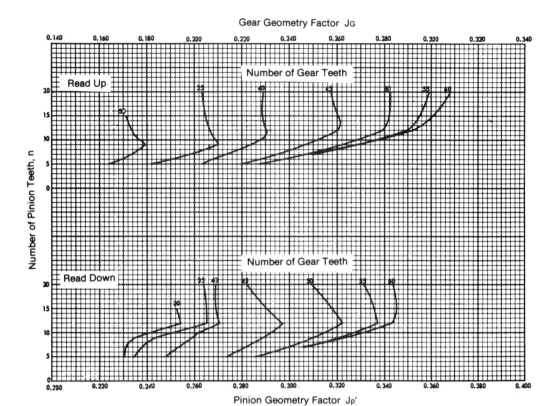

FIG. 3-47 Geometry factor J for strength of hypoid gears with 19° average pressure angle and E/D ratio of 0.10.

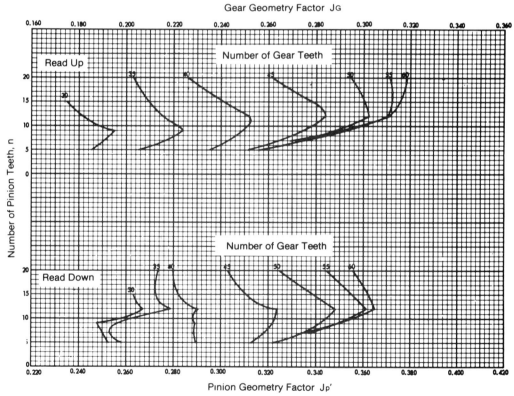

FIG. 3-48 Geometry factor J for strength of hypoid gears with 19° average pressure angle and E/D ratio of 0.15.

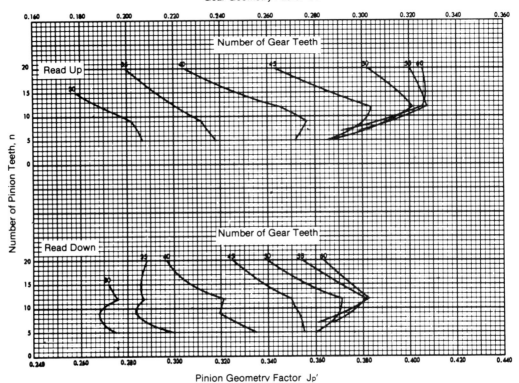

FIG. 3-49 Geometry factor J for strength of hypoid gears with 19° average pressure angle and E/D ratio of 0.20.

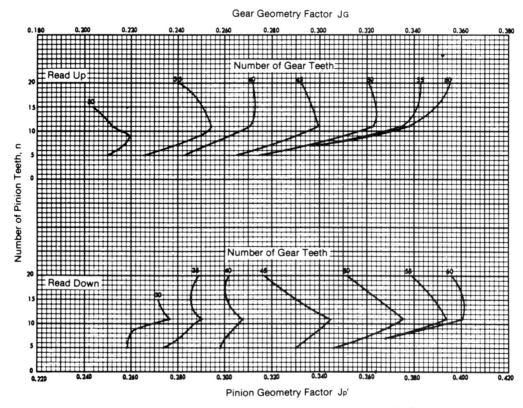

FIG. 3-50 Geometry factor J for strength of hypoid gears with $22\frac{1}{2}°$ average pressure angle and E/D ratio of 0.10.

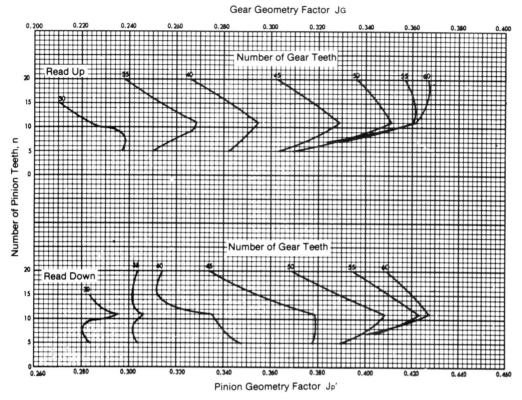

FIG. 3-51 Geometry factor J for strength of hypoid gears with $22\frac{1}{2}°$ average pressure angle and E/D ratio of 0.15.

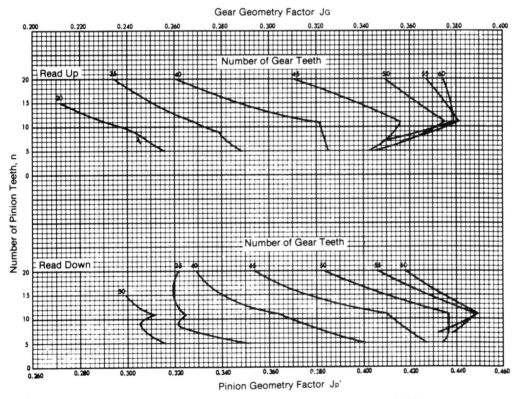

FIG. 3-52 Geometry factor J for strength of hypoid gears with $22\frac{1}{2}°$ average pressure angle and E/D ratio of 0.20.

TABLE 3-13 Allowable Contact Stress S_{ac}

Material	Heat treatment	Minimum hardness		Contact stress S_{ac}, lb/in²
		Brinell	Rockwell C	
Steel	Carburized (case-hardened)		60	250 000
Steel	Carburized (case-hardened)		55	210 000
Steel	Flame- or induction-hardened	500	50	200 000
Steel and nodular iron	Hardened and tempered	400		180 000
Steel	Nitrided		60	180 000
Steel and nodular iron	Hardened and tempered	300		140 000
Steel and nodular iron	Hardened and tempered	180		100 000
Cast iron	As cast	200		80 000
Cast iron	As cast	175		70 000
Cast iron	As cast			60 000

TABLE 3-14 Allowable Bending Stress S_{at}

Material	Heat treatment	Surface hardness		Bending stress S_{at}, lb/in²
		Brinell	Rockwell C	
Steel	Carburized (case-hardened)	575–625	55 min.	60 000
Steel	Flame- or induction-hardened (unhardened root fillet)	450–500	50 min.	27 000
Steel	Hardened and tempered	450 min.		50 000
Steel	Hardened and tempered	300 min.		42 000
Steel	Hardened and tempered	180 min.		28 000
Steel	Normalized	140 min.		22 000
Cast iron	As cast	200 min.		13 000
Cast iron	As cast	175 min.		8 500
Cast iron	As cast			5 000

Notes · Drawings · Ideas

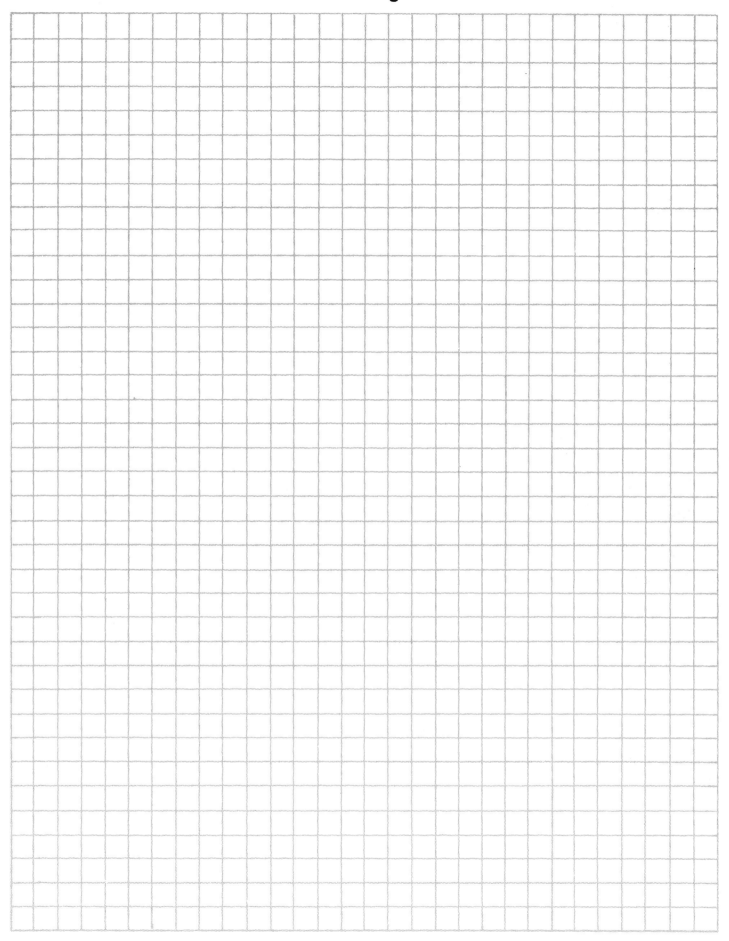

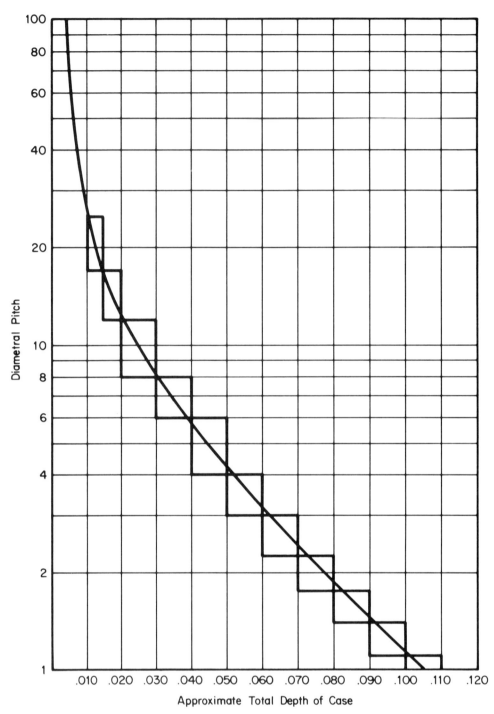

FIG. 3-53 Diametral pitch versus total case depth. In case of choice, use the greater case depth on ground gears or on short face widths.

Notes • Drawings • Ideas

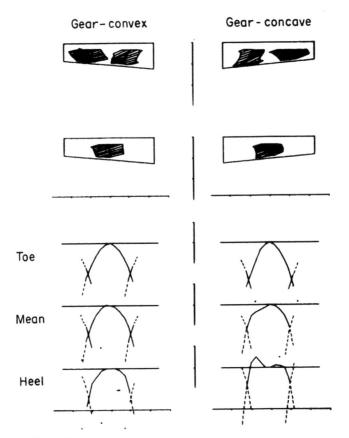

FIG. 3-54 Typical tooth contact analysis graph.

chapter 4
HELICAL GEARS

RAYMOND J. DRAGO, P.E.
Senior Engineer, Advanced Power Train Technology
Boeing Vertol Company
Philadelphia, Pennsylvania

The following is quoted from the Foreword of Ref. [4-1]:

> This AGMA Standard and related publications are based on typical or average data, conditions, or applications. The standards are subject to continual improvement, revision, or withdrawal as dictated by increased experience. Any person who refers to AGMA technical publications should be sure that he has the latest information available from the Association on the subject matter.
>
> Tables or other self-supporting sections may be quoted or extracted in their entirety. Credit line should read: "Extracted from AGMA Standard for Rating the Pitting Resistance and Bending Strength of Spur and Helical Involute Gear Teeth, AGMA 218.01, with the permission of the publisher, American Gear Manufacturers Association, 101 South Peyton Street, Alexandria, Virginia 22314."

This reference is cited because numerous American Gear Manufacturer's Association (AGMA) tables and figures are used in this chapter. In each case the appropriate publication is noted in a footnote or figure caption.

4-1 INTRODUCTION

Helical gearing, in which the teeth are cut at an angle with respect to the axis of rotation, is a later development than spur gearing and has the advantage that the action is smoother and tends to be quieter. In addition, the load transmitted may be somewhat larger, or the life of the gears may be greater for the same loading than with an equivalent pair of spur gears. Helical gears produce an end thrust along the axis of the shafts in addition to the separating and tangential (driving) loads of spur gears. Where suitable means can be provided to take this thrust, such as thrust collars or ball or tapered-roller bearings, it is no great disadvantage.

Conceptually, helical gears may be thought of as stepped spur gears in which the size of the step becomes infinitely small. For external parallel-axis helical gears to mesh, they must have the same helix angle but be of different hand. An external-internal set will, however, have equal helix angle with the same hand.

Involute profiles are usually employed for helical gears, and the same comments made earlier about spur gears hold true for helical gears.

Although helical gears are most often used in a parallel-axis arrangement, they can also be mounted on nonparallel noncoplanar axes. Under such mounting conditions they will, however, have limited load capacity.

Although helical gears which are used on crossed axes are identical in geometry and manufacture to those used in parallel axes, their operational characteristics are quite different. For this reason they are discussed separately at the end of this chapter. All the forthcoming discussion therefore applies only to helical gears operating on parallel axes.

4-2 TYPES

Helical gears may take several forms, as shown in Fig. 4-1:

1. Single
2. Double conventional
3. Double staggered
4. Continuous (herringbone)

Single-helix gears are readily manufactured on conventional gear cutting and grinding equipment. If the space between the two rows of a double-helix gear is wide enough, they, too, may be cut and ground, if necessary, on conventional equipment. Continuous or herringbone gears, however, can be cut only on a special shaping machine (Sykes) and usually cannot be ground at all.

Only single-helix gears may be used in a crossed-axis configuration.

4-3 ADVANTAGES

There are three main reasons why helical rather than straight spur gears are used in a typical application. These are concerned with the noise level, the load capacity, and the manufacturing.

4-3-1 Noise

Helical gears produce less noise than spur gears of equivalent quality because the total contact ratio is increased. Figure 4-2 shows this effect quite dramatically. However, these results are measured at the mesh for a specific test setup; thus, although the trend is accurate, the absolute results are not.

Figure 4-2 also brings out another interesting point. At high values of helix angle, the improvement in noise tends to peak; that is, the curve flattens out. Had data been obtained at still higher levels, the curve would probably drop drastically. This is due to the difficulty in manufacturing and mounting such gears accurately enough to take full advantage of the improvement in contact ratio. These effects at very high helix angles actually tend to reduce the effective contact ratio, and so noise increases. Since helix angles greater than 45° are seldom used and are generally impractical to manufacture, this phenomenon is of academic interest only.

4-3-2 Load Capacity

As a result of the increased total area of tooth contact available, the load capacity of helical gears is generally higher than that for equivalent spur gears. The reason for this increase is obvious when we consider the contact line comparison which Fig. 4-3 shows. The most critical load condition for a spur gear occurs when a single

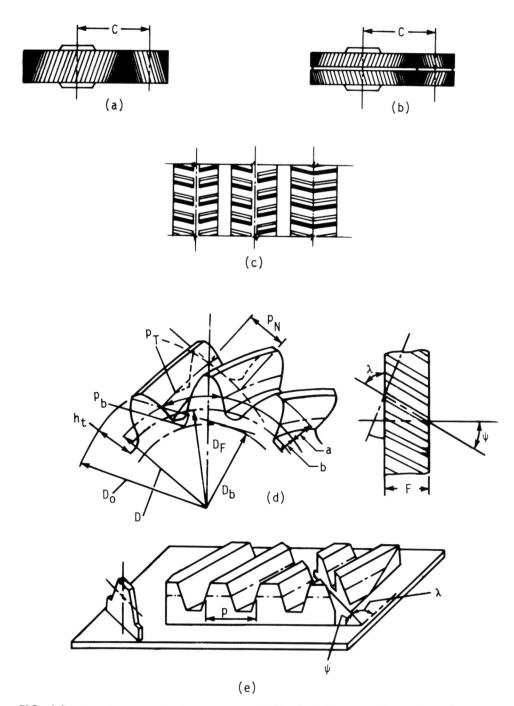

FIG. 4-1 Terminology of helical gearing. (*a*) Single-helix gear. (*b*) Double-helix gear. (*c*) Types of double-helix gears, left, conventional; center, staggered; right, continuous or herringbone. (*d*) Geometry. (*e*) Helical rack.

tooth carries all the load at the highest point of single-tooth contact (Fig. 4-3c). In this case the total length of the contact line is equal to the face width. In a helical gear, since the contact lines are inclined to the tooth with respect to the face width, the total length of the line of contact is increased (Fig. 4-3b) so that it is greater than the face width. This lowers unit loading and thus increases capacity.

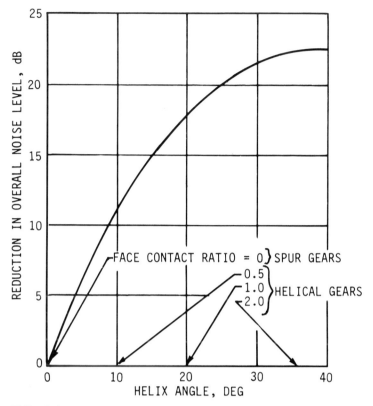

FIG. 4-2 Effect of face-contact ratio on noise level. Note that increased helix angles lower the noise level.

4-3-3 Manufacturing

In the design of a gear system, it is often necessary to use a specific ratio on a specific center distance. Frequently this results in a diametral pitch which is nonstandard. if helical gears are employed, a limited number of standard cutters may be used to cut a wide variety of transverse-pitch gears, simply by varying the helix angle, thus allowing virtually any center-distance and tooth number combination to be accommodated.

4-4 GEOMETRY

When they are considered in the transverse plan (that is, a plane perpendicular to the axis of the gear), all helical-gear geometry is identical to that for spur gears. Standard tooth proportions are usually based on the normal diametral pitch, as shown in Table 4-1.

It is frequently necessary to convert from the normal plane to the transverse plane and vice versa. Table 4-2 gives the necessary equations. All calculations previously defined for spur gears with respect to transverse or profile-contact ratio, top land, lowest point of contact, true involute form radius, nonstandard center, etc., are valid for helical gears if only a transverse plane section is considered.

For spur gears the profile-contact ratio (ratio of contact to the base pitch) must be

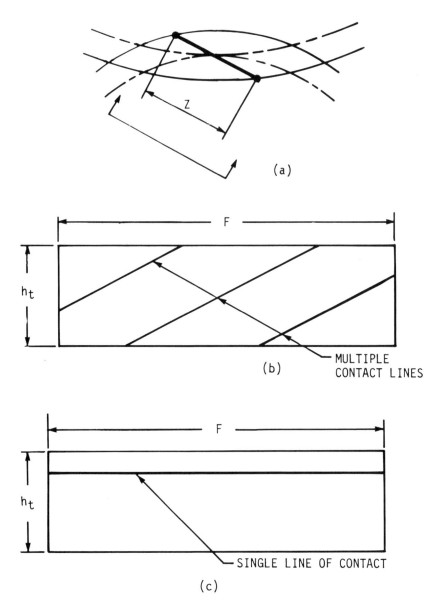

FIG. 4-3 Comparison of spur and helical contact lines. (*a*) Transverse section; (*b*) helical contact lines; (*c*) spur contact line.

greater than unity for uniform rotary-motion transmission to occur. Helical gears, however, provide an additional overlap along the axial direction; thus their profile-contact ratio need not necessarily be greater than unity. The sum of both the profile-contact ratio and the axial overlap must, however, be at least unity. The axial overlap, also often called the *face-contact ratio,* is the ratio of the face width to the axial pitch. The face-contact ratio is given by

$$m_F = \frac{P_{do} F \tan \psi_o}{\pi} \qquad (4\text{-}1)$$

where P_{do} = operating transverse diametral pitch
ψ_o = helix angle at operating pitch circle
F = face width

TABLE 4-1 Standard Tooth Proportions for Helical Gears

Quantity†	Formula	Quantity†	Formula
Addendum	$\dfrac{1.00}{P_N}$	External gears:	
Dedendum	$\dfrac{1.25}{P_N}$	Standard center distance	$\dfrac{D+d}{2}$
Pinion pitch diameter	$\dfrac{N_P}{P_N \cos \psi}$	Gear outside diameter	$D + 2a$
Gear pitch diameter	$\dfrac{N_G}{P_N \cos \psi}$	Pinion outside diameter	$d + 2a$
Normal arc tooth thickness	$\dfrac{\pi}{P_N} - \dfrac{B_N}{2}$	Gear root diameter	$D - 2b$
Pinion base diameter	$d \cos \phi_T$	Pinion root diameter	$d - 2b$
Gear base diameter	$D \cos \phi_T$	Internal gears: Center distance	$\dfrac{D-d}{2}$
Base helix angle	$\tan^{-1}(\tan \psi \cos \phi_T)$	Inside diameter Root diameter	$d - 2a$ $D + 2b$

†All dimensions in inches, and angles are in degrees.

Other parameters of interest in the design and analysis of helical gears are the base pitch p_b and the length of the line of action Z, both in the transverse plane. These are

$$p_b = \frac{\pi}{P_d} \cos \phi_T \qquad (4\text{-}2)$$

and

$$Z = (r_o^2 - r_b^2)^{1/2} + (R_o^2 - R_b^2)^{1/2} - C_o \sin \phi_o \qquad (4\text{-}3)$$

TABLE 4-2 Conversions between Normal and Transverse Planes

Parameter (normal/transverse)	Normal to transverse	Transverse to normal
Pressure angle (ϕ_N/ϕ_T)	$\phi_T = \tan^{-1} \dfrac{\tan \phi_N}{\cos \psi}$	$\phi_N = \tan^{-1}(\tan \phi_T \cos \psi)$
Diametral pitch (P_N/P_d)	$P_d = P_N \cos \psi$	$P_N = \dfrac{P_d}{\cos \psi}$
Circular pitch (p_N/p_T)	$P_T = \dfrac{P_N}{\cos \psi}$	$P_N = P_T \cos \psi$
Arc tooth thickness (T_N/T_T)	$T_T = \dfrac{T_N}{\cos \psi}$	$T_N = T_T \cos \psi$
Backlash (B_N/B_T)	$B_T = \dfrac{B_N}{\cos \psi}$	$B_N = B_T \cos \psi$

This equation is for an external gear mesh. For an internal gear mesh, the length of the line of action is

$$Z = (R_I^2 - R_b^2)^{1/2} - (r_o^2 - r_b^2)^{1/2} + C_o \sin \phi_o \qquad (4\text{-}4)$$

where P_d = transverse diametral pitch as manufactured
ϕ_T = transverse pressure angle as manufactured, degrees (deg)
r_o = effective pinion outside radius, inches (in)
R_o = effective gear outside radius, in
R_I = effective gear inside radius, in
ϕ_o = operating transverse pressure angle, deg
r_b = pinion base radius, in
R_b = gear base radius, in
C_o = operating center distance, in

The operating transverse pressure angle ϕ_o is

$$\phi_o = \cos^{-1}\left(\frac{C}{C_o} \cos \phi_T\right) \qquad (4\text{-}5)$$

The manufactured center distance C is simply

$$C = \frac{N_P + N_G}{2P_d} \qquad (4\text{-}6)$$

for external mesh; for internal mesh the relation is

$$C = \frac{N_G - N_P}{2P_d} \qquad (4\text{-}7)$$

The contact ratio m_P in the transverse plane (profile-contact ratio) is defined as the ratio of the total length of the line of action in the transverse plane Z to the base pitch in the transverse plane p_b. Thus

$$m_P = \frac{Z}{P_b} \qquad (4\text{-}8)$$

The diametral pitch, pitch diameters, helix angle, and the normal pressure angle at the operating pitch circle are required in the load-capacity evaluation of helical gears. These terms are given by

$$P_{do} = \frac{N_P + N_G}{2C_o} \qquad (4\text{-}9)$$

for external mesh; for internal mesh

$$P_{do} = \frac{N_G - N_P}{2C_o} \qquad (4\text{-}10)$$

Also

$$d = \frac{N_P}{P_{do}} \qquad D = \frac{N_G}{P_{do}} \qquad (4\text{-}11)$$

$$\psi_B = \tan^{-1}(\tan \psi \cos \phi_T) \qquad (4\text{-}12)$$

Notes · Drawings · Ideas

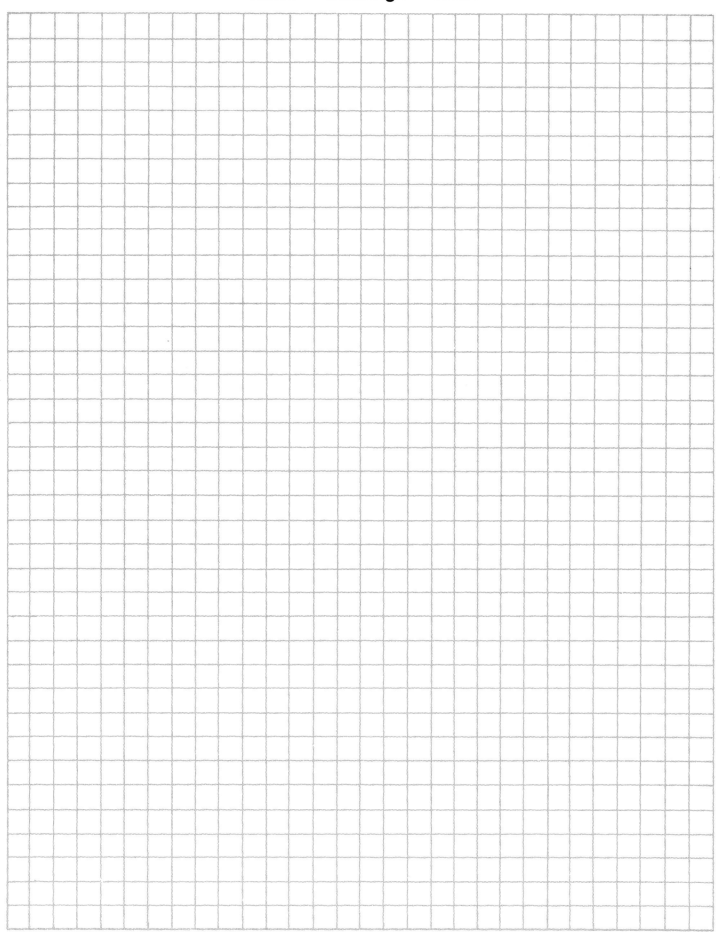

$$\psi_o = \tan^{-1} \frac{\tan \psi_B}{\cos \phi_o} \qquad (4\text{-}13)$$

$$\phi_{No} = \sin^{-1}(\sin \phi_o \cos \psi_B) \qquad (4\text{-}14)$$

where P_{do} = operating diametral pitch
ψ_B = base helix angle, deg
ψ_o = helix angle at operating pitch point, deg
ϕ_{No} = operating normal pressure angle, deg
d = operating pinion pitch diameter, in
D = operating gear pitch diameter, in

4-5 LOAD RATING

Reference [4-1] establishes a coherent method for rating external helical and spur gears. The treatment of strength and durability provided here is derived in large part from this source.

Four factors must be considered in the load rating of a helical-gear set: strength, durability, wear resistance, and scoring probability. Although strength and durability must always be considered, wear resistance and scoring evaluations may not be required for every case. We treat each topic in some depth.

4-5-1 Strength and Durability

The strength of a gear tooth is evaluated by calculating the bending stress index number at the root by

$$s_t = \frac{W_t K_a}{K_v} \frac{P_d}{F_E} \frac{K_b K_m}{J} \qquad (4\text{-}15)$$

where s_t = bending stress index number, pounds per square inch (psi)
K_a = bending application factor
F_E = effective face width, in
K_m = bending load-distribution factor
K_v = bending dynamic factor
J = bending geometry factor
P_d = transverse operating diametral pitch
K_b = rim thickness factor

The calculated bending stress index number s_t must be within safe operating limits as defined by

$$s_t \leq \frac{s_{at} K_L}{K_T K_R} \qquad (4\text{-}16)$$

where s_{at} = allowable bending stress index number
K_L = life factor
K_T = temperature factor
K_R = reliability factor

Notes · Drawings · Ideas

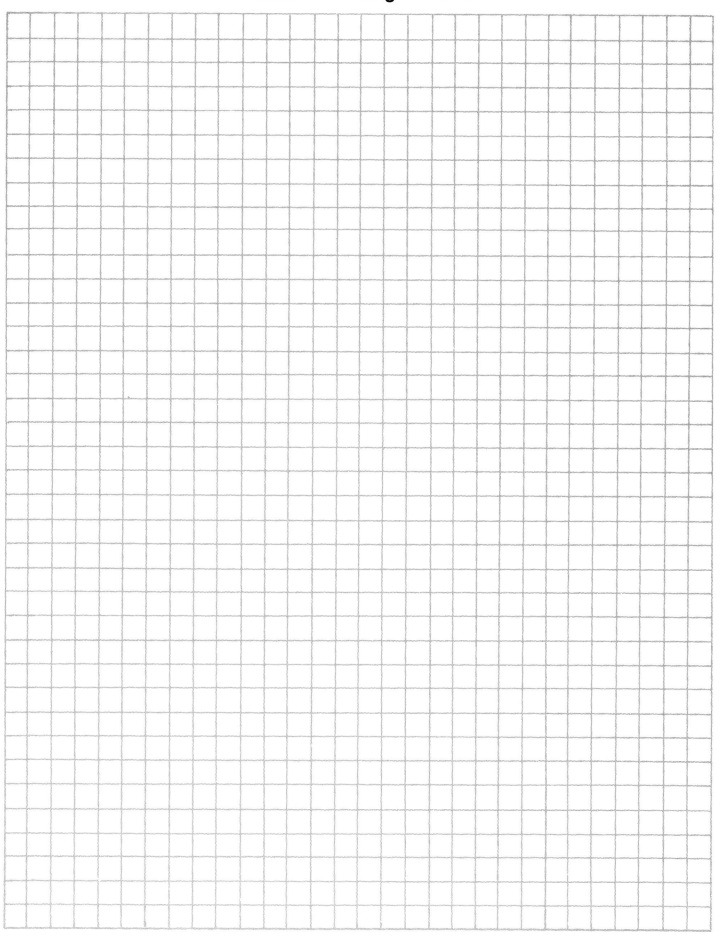

Some of the factors which are used in these equations are similar to those used in the durability equations. Thus we present the basic durability rating equations before discussing the factors:

$$s_c = C_P \sqrt{\frac{W_t C_a}{C_v} \frac{1}{dF_N} \frac{C_m}{I}} \quad (4\text{-}17)$$

where s_c = contact stress index number
C_a = durability application factor
C_v = durability dynamic factor
d = operating pinion pitch diameter
F_N = net face width, in
C_m = load-distribution factor
C_p = elastic coefficient
I = durability geometry factor

The calculated contact stress index number must be within safe operating limits as defined by

$$s_c \leq \frac{s_{ac} C_L C_H}{C_T C_R} \quad (4\text{-}18)$$

where s_{ac} = allowable contact stress index number
C_L = durability life factor
C_H = hardness ratio factor
C_T = temperature factor
C_R = reliability factor

To utilize these equations, each factor must be evaluated. The tangential load W_t is given by

$$W_t = \frac{2T_P}{d} \quad (4\text{-}19)$$

where T_P = pinion torque in inch-pounds (in·lb) and d = pinion operating pitch diameter in inches. If the duty cycle is not uniform but does not vary substantially, then the maximum anticipated load should be used. Similarly, if the gear set is to operate at a combination of very high and very low loads, it should be evaluated at the maximum load. If, however, the loading varies over a well-defined range, then the cumulative fatigue damage for the loading cycle should be evaluated by using Miner's rule. For a good explanation see Ref. [4-2].

APPLICATION FACTORS C_a AND K_a. This factor makes the allowances for externally applied loads of unknown nature which are in excess of the nominal tangential load. Such factors can be defined only after considerable field experience has been established. In a *new* design this consideration places the designer squarely on the horns of a dilemma, since "new" presupposes limited, if any, experience. The values shown in Table 4-3 may be used as a guide if no other basis is available.

The application factor should never be set equal to unity except where clear experimental evidence indicates that the loading will be absolutely uniform. Wherever possible, the actual loading to be applied to the system should be defined. One of the most common mistakes made by gear system designers is assuming that the motor (or engine, etc.) "nameplate" rating is also the gear unit rating point.

Notes · Drawings · Ideas

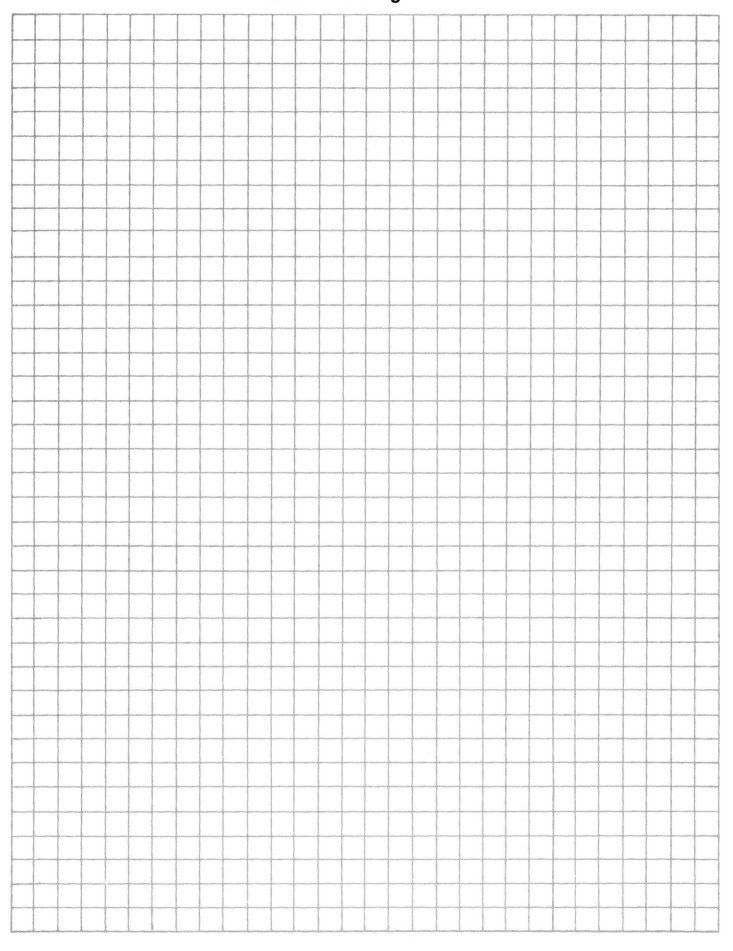

TABLE 4-3 Application Factor Guidelines

Power source	Character of load on driven machine		
	Uniform	Moderate shock	Heavy shock
Uniform	1.15	1.25	At least 1.75
Light shock	1.25	1.50	At least 2.00
Medium shock	1.50	1.75	At least 2.50

DYNAMIC FACTORS C_v AND K_v. These factors account for internally generated tooth loads which are induced by nonconjugate meshing action. This discontinuous motion occurs as a result of various tooth errors (such as spacing, profile, and runout) and system effects (such as deflections). Other effects, such as system torsional resonances and gear blank resonant responses, may also contribute to the overall dynamic loading experienced by the teeth. The latter effects must, however, be separately evaluated. The effect of tooth accuracy may be determined from Fig. 4-4, which is based on both pitch line velocity and gear quality Q_n as specified in Ref. [4-3]. The pitch line velocity of a gear is

$$v_t = 0.2618 nD \qquad (4\text{-}20)$$

where v_t = pitch line velocity, feet per minute (fpm)
n = gear speed, revolutions per minute (rpm)
D = gear pitch diameter, in

EFFECTIVE AND NET FACE WIDTHS F_E AND F_N. The net minimum face width of the narrowest member should always be used for F_N. In cases where one member

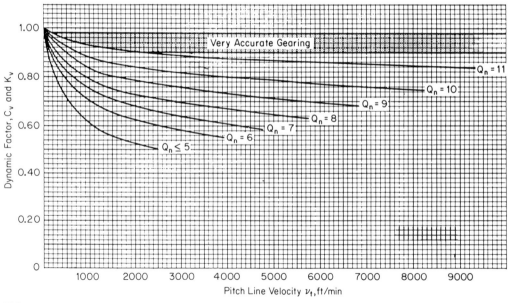

FIG. 4-4 Dynamic factors C_v and K_v. *(From Ref. [4-1].)*

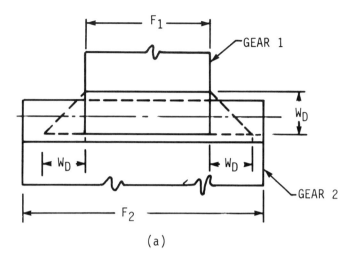

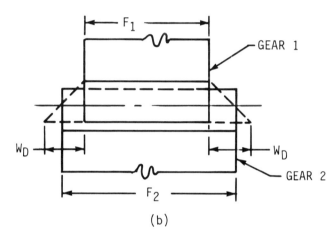

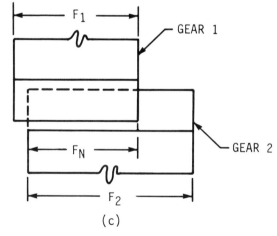

FIG. 4-5 Definition of effective face width. (a) $F_{E1} = F_1$, $F_{E2} = F_1 + 2W_D$; $F_N = F_1$; (b) $F_{E1} = F_1$, $F_{E2} = F_2$, $F_N = F_1$; (c) $F_{E1} = F_{E2} = F_N$.

has a substantially larger face width than its mate, some advantage may be taken of this fact in the bending stress calculations, but it is unlikely that a very narrow tooth will fully transfer its tooth load across the face width of a much wider gear. At best, the effective face width of a larger-face-width gear mating with a smaller-face-width gear is limited to the minimum face of the smaller member plus some allowance for the extra support provided by the wide face. Figure 4-5 illustrates the definition of net and effective face widths for various cases.

RIM THICKNESS FACTOR K_b. The basic bending stress equations were developed for a single tooth mounted on a rigid support so that it behaves as a short cantilever beam. As the rim which supports the gear tooth becomes thinner, a point is reached at which the rim no longer provides "rigid" support. When this occurs, the bending of the rim itself combines with the tooth bending to yield higher total alternating stresses than would be predicted by the normal equations. Additionally, when a tooth is subjected to fully reversed bending loads, the alternating stress is also increased because of the additive effect of the compressive stress distribution on the normally unloaded side of the tooth, as Fig. 4-6 shows. Both effects are accounted for by the rim thickness factor, as Fig. 4-7 indicates.

It must be emphasized that the data shown in Fig. 4-7 are based on a limited amount of analytical and experimental (photoelastic and strain-gauge) measurements and thus must be used judiciously. Still, they are the best data available to date and are far better than nothing at all; see Refs. [4-4] and [4-5].

For gear blanks which utilize a T-shaped rim and web construction, the web acts as a hard point, if the rim is thin, and stresses will be higher over the web than over the ends of the T. The actual value which should be used for such constructions depends greatly on the relative proportions of the gear face width and the web. If the web spans 70 to 80 percent of the face width, the gear may be considered as having a rigid backup. Thus the backup ratio will be greater than 2.0, and any of the curves shown may be used (that is, curve C or B, both of which are identical above a 2.0 backup ratio, for unidirectional loading or curve A for fully reversed loading). If the proportions are between these limits, the gear lies in a gray area and probably lies somewhere in the range defined by curves B and C. Some designer discretion should be exercised here.

Finally, note that the rim thickness factor is equal to unity only for unidirectionally loaded, rigid-backup helical gears. For fully reversed loading, its value will be at least 1.4, even if the backup is rigid.

LOAD-DISTRIBUTION FACTORS K_m AND C_c. These factors modify the rating equations to account for the manner in which the load is distributed on the teeth. The load on a set of gears will never be exactly uniformly distributed. Factors which effect the load distribution include the accuracy of the teeth themselves; the accuracy of the housing which supports the teeth (as it influences the alignment of the gear axes); the deflections of the housing, shafts, and gear blanks (both elastic and thermal); and the internal clearances in the bearings which support the gears among others.

All these and any other appropriate effects must be evaluated in order to define the total effective alignment error e_t for the gear pair. Once this is accomplished, the load-distribution factor may be calculated.

In some cases it may not be possible to fully define or even estimate the value of e_t. In such cases an empirical approach may be used. We discuss both approaches in some detail.

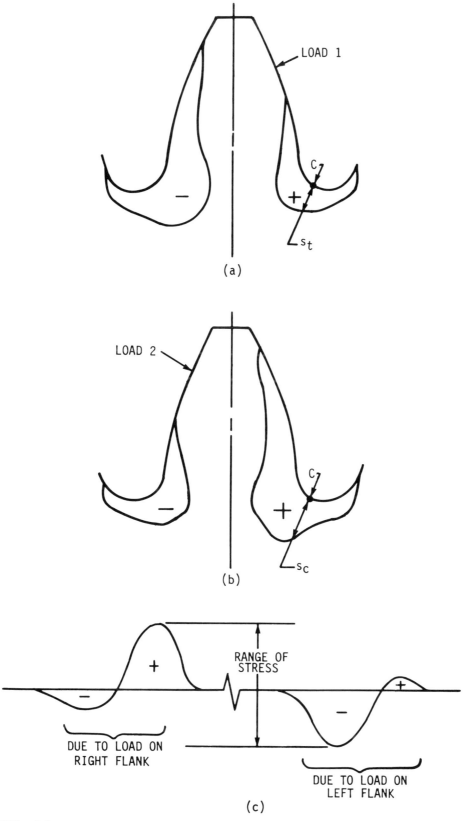

FIG. 4-6 Stress condition for reversing (as with an idler) loading. (*a*) Load on right flank; (*b*) load on left flank; (*c*) typical waveform for strain gauge at point *C*.

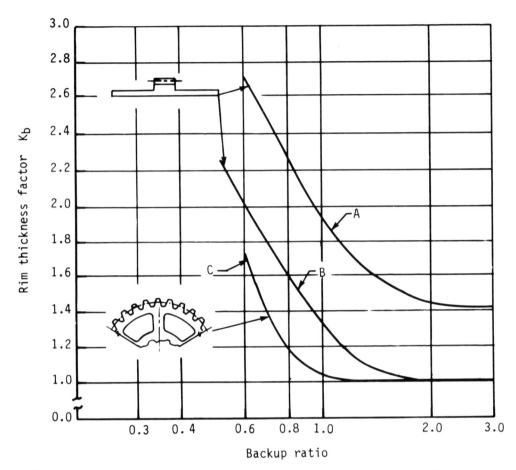

FIG. 4-7 Rim thickness factor K_b. The *backup ratio* is defined as the ratio of the rim thickness to the tooth height. Curve A is fully reversed loading; curves B and C are unidirectional loading.

The empirical approach requires only minimal data, and so it is the simplest to apply. Several conditions must be met, however, prior to using this method:

1. Net face width to pinion pitch diameter ratios must be less than or equal to 2.0. (For double-helix gears the gap is not included in the face width.)
2. The gear elements are mounted between bearings (not overhung).
3. Face width can be up to 40 in.
4. There must be contact across the full face width of the narrowest member when loaded.
5. Gears are not highly crowned.

The empirical expression for the load-distribution factor is

$$C_m = K_m = 1.0 + C_{mc}(C_{pf}C_{pm} + C_{ma}C_e) \qquad (4\text{-}21)$$

where C_{mc} = lead correction factor
C_{pf} = pinion proportion factor
C_{pm} = pinion proportion modifier

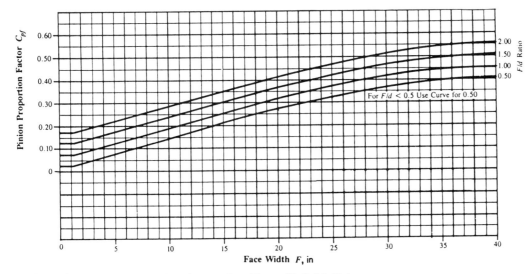

FIG. 4-8 Pinion proportion factor C_{pf}. *(From Ref. [4-1].)*

C_{ma} = mesh alignment factor
C_e = mesh alignment correction factor

The lead correction factor C_{mc} modifies the peak loading in the presence of slight crowning or lead correction as follows:

$$C_{mc} = \begin{cases} 1.0 & \text{for gear with unmodified leads} \\ 0.8 & \text{for gear with leads properly modified by crowning or lead correction} \end{cases}$$

Figure 4-8 shows the pinion proportion factor C_{pf} which accounts for deflections due to load. The pinion proportion modifier C_{pm} alters C_{pf} based on the location of the pinion relative to the supporting bearings. Figure 4-9 defines the factors S and S_1. And C_{pm} is defined as follows:

$$C_{pm} = \begin{cases} 1.0 & \text{when } S_1/S < 0.175 \\ 1.1 & \text{when } S_1/S \geq 0.175 \end{cases}$$

The mesh alignment factor C_{ma} accounts for factors other than elastic deformations. Figure 4-10 provides values for this factor for four accuracy groupings. For double-helix gears this figure should be used with F equal to half of the total face width. The mesh alignment correction factor C_e modifies the mesh alignment factor to allow for the improved alignment which may be obtained when a gear set is adjusted at assembly or when the gears are modified by grinding, skiving, or lapping to more closely match their mates at assembly (in which case, pinion and gear become a matched set). Only two values are permissible for C_e—either 1.0 or 0.8, as defined by the following requirements:

$$C_e = \begin{cases} 0.80 & \text{when the compatibility of the gearing is improved by lapping, grinding, or skiving after trial assembly to improve contact} \\ 0.80 & \text{when gearing is adjusted at assembly by shimming support bearings and/or housing to yield uniform contact} \\ 1.0 & \text{for all other conditions} \end{cases}$$

126 GEARING: A MECHANICAL DESIGNERS' WORKBOOK

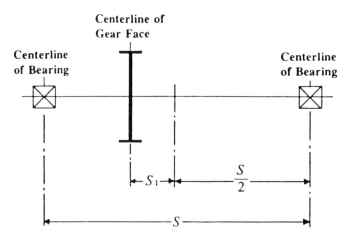

FIG. 4-9 Definition of distances S and S_1. Bearing span is distance S; pinion offset from midspan is S_1. *(From Ref. [4-1].)*

If enough detailed information is available, a better estimate of the load-distribution factor may be obtained by using a more analytical approach. This method, however, requires that the total alignment error e_t be calculated or estimated. Depending on the contact conditions, one of two expressions are used to calculate the load-distribution factor.

If the tooth contact pattern at normal operating load essentially covers the entire available tooth face, Eq. (4-22) should be used. If the tooth contact pattern does not

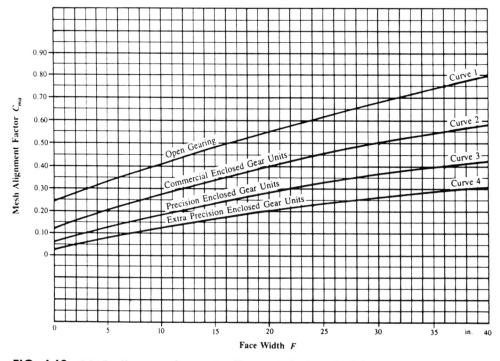

FIG. 4-10 Mesh alignment factor C_{ma}. For analytical method for determination of C_{ma} see Eq. (4-21). *(From Ref. [4-1].)*

Notes · Drawings · Ideas

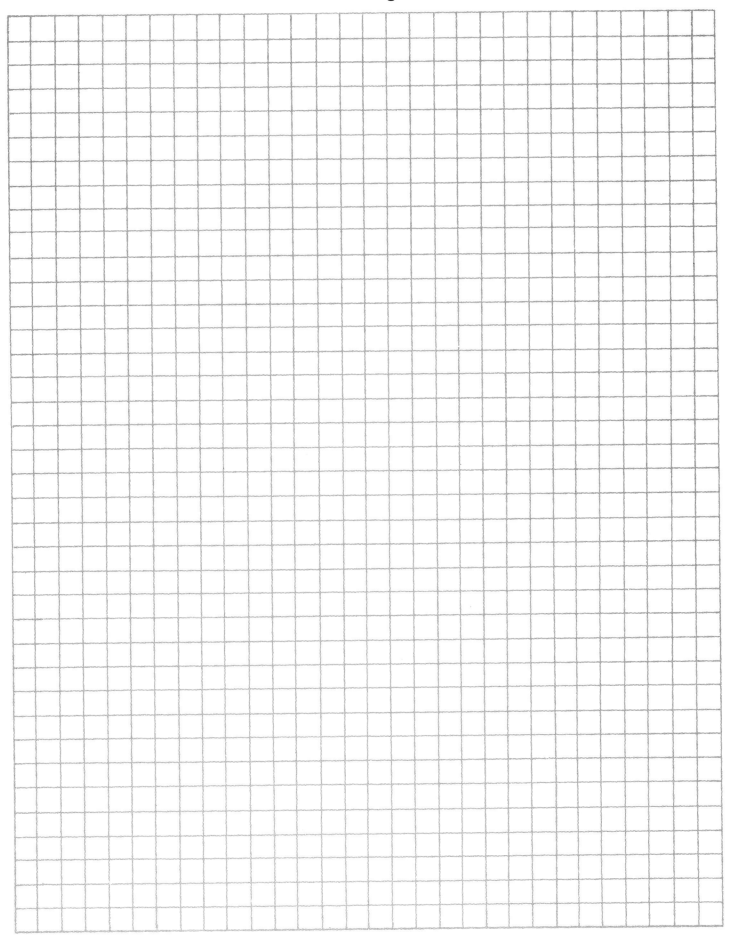

cover the entire available tooth face (as would be the case for poorly aligned or high-crowned gears) at normal operating loads, then Eq. (4-23) must be used:

$$C_m = 1.0 + \frac{Ge_tZF}{4W_t p_b} \qquad (4\text{-}22)$$

and

$$C_m = \sqrt{\frac{Ge_tZF}{W_t P_b}} \qquad (4\text{-}23)$$

where W_t = tangential tooth load, pounds (lb)
G = tooth stiffness constant, (lb/in)/in of face
Z = length of line of contact in transverse plane
e_t = total effective alignment error, in/in
p_b = transverse base pitch, in
F = net face width of narrowest member, in

The value of G will vary with tooth proportions, tooth thickness, and material. For steel gears of standard or close to standard proportions, it is normally in the range of 1.5×10^6 to 2.0×10^6 psi. The higher value should be used for higher-pressure-angle teeth which are normally stiffer, while the lower value is representative of more flexible teeth. The most conservative approach is to use the higher value in all cases.

For double-helix gears, each half should be analyzed separately by using the appropriate values of F and e_t and by assuming half of the tangential tooth load is transmitted by each half (the values for p_b, Z, and G remain unchanged).

GEOMETRY FACTOR I. The geometry factor I evaluates the radii of curvature of the contacting tooth profiles based on the pressure angle, helix, and gear ratio. Effects of modified tooth proportions and load sharing are considered. The I factor is defined as follows:

$$I = \frac{C_c C_x C_\psi^2}{m_N} \qquad (4\text{-}24)$$

where C_c = curvature factor at operating pitch line
C_x = contact height factor
C_ψ = helical overlap factor
m_N = load-sharing ratio

The curvature factor is

$$C_c = \frac{\cos \phi_o \sin \phi_o}{2} \frac{N_G}{N_G + N_P} \qquad (4\text{-}25)$$

for external mesh; for internal mesh

$$C_c = \frac{\cos \phi_o \sin \phi_o}{2} \frac{N_G}{N_G - N_P} \qquad (4\text{-}26)$$

The contact height factor C_x adjusts the location on the tooth profile at which the critical contact stress occurs (i.e., face-contact ratio > 1.0). The stress is calculated

Notes · Drawings · Ideas

at the mean diameter or the middle of the tooth profile. For low-contact-ratio helical gears (that is, face contact ratio ≤ 1.0), the stress is calculated at the lowest point of single-tooth contact in the transverse plane and C_x is given by Eq. (4-27):

$$C_x = \frac{R_1 R_2}{R_P R_G} \tag{4-27}$$

where R_P = pinion curvature radius at operating pitch point, in
R_G = gear curvature radius at operating pitch point, in
R_1 = pinion curvature radius at critical contact point, in
R_2 = gear curvature radius at critical contact point, in

The required radii are given by

$$R_P = \frac{d}{2} \sin \phi_o \qquad R_G = \frac{D}{2} \sin \phi_o \tag{4-28}$$

where d = pinion operating pitch diameter, in
D = gear operating pitch diameter, in
ϕ_o = operating pressure angle in transverse plane, deg

$$R_1 = R_P - Z_c \tag{4-29}$$

and
$$R_2 = R_G + Z_c \tag{4-30}$$

for external gears; for internal gears

$$R_2 = R_G - Z_c \tag{4-31}$$

where Z_c is the distance along the line of action in the transverse plane to the critical contact point. The value of Z_c is dependent on the transverse contact ratio. For helical gears, where the face-contact ratio ≤ 1.0, Z_c is found by using Eq. (4-32). For normal helical gears where the face-contact ratio is > 1.0, Eq. (4-33) is used:

$$Z_c = p_b - 0.5[(d_o^2 - d_b^2)^{1/2} - (d^2 - d_b^2)^{1/2}] \quad m_F \leq 1.0 \tag{4-32}$$

and
$$Z_c = 0.5\,[(d^2 - d_b^2)^{1/2} - (d_m^2 - d_b^2)^{1/2}] \quad m_F > 1.0 \tag{4-33}$$

where p_b = base pitch, in
d_o = pinion outside diameter, in
d_b = pinion base diameter, in
d_m = pinion mean diameter, in

The pinion mean diameter is defined by Eq. (4-34) or (4-35). For external mesh,

$$d_m = C_o - \frac{D_o - d_o}{2} \tag{4-34}$$

For internal mesh,

$$d_m = \frac{D_I + d_o}{2} - C_o \tag{4-35}$$

Notes · Drawings · Ideas

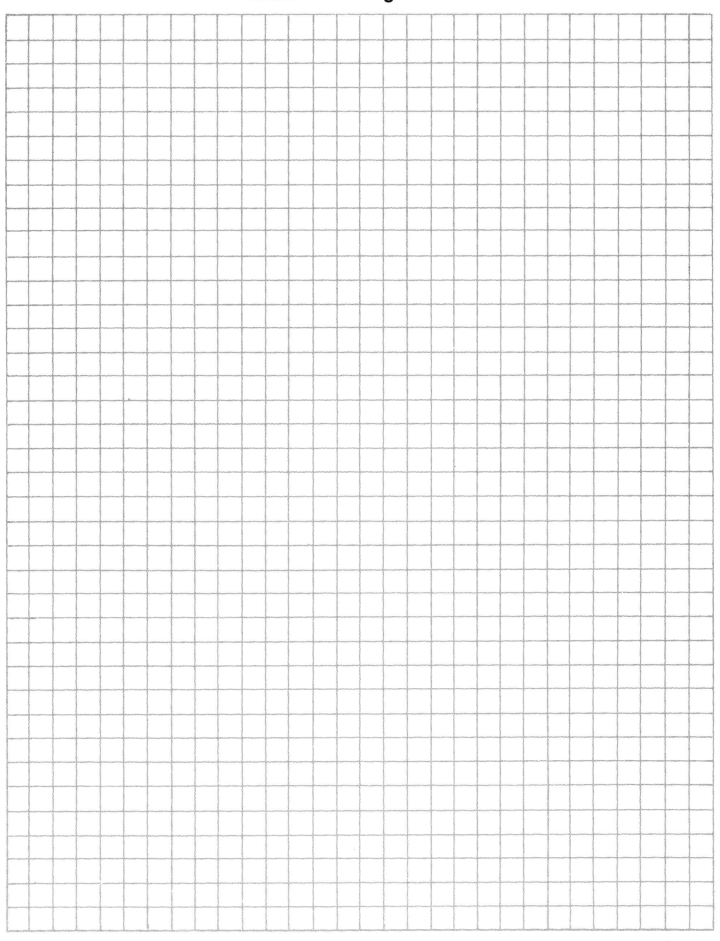

where D_o = external gear outside diameter and D_I = internal gear inside diameter.

The helical factor C_ψ accounts for the partial helical overlap action which occurs in helical gears with a face-contact ratio $m_F \leq 1.0$. For helical gears with a face-contact ratio > 1.0, C_ψ is set equal to unity: for low-contact helical gears it is

$$C_\psi = \sqrt{1 - m_F + \frac{C_{xn} Z m_F^2}{C_x F \sin \psi_b}} \tag{4-36}$$

where Z = total length of line of action in transverse plane, in
F = net minimum face width, in
m_F = face-contact ratio
C_x = contact height factor [Eq. (4-27)]
C_{xn} = contact height factor for equivalent normal helical gears [Eq. (4-37)]
ψ_b = base helix angle, deg

The C_{xn} factor is given by

$$C_{xn} = \frac{R_{1n} R_{2n}}{R_P R_G} \tag{4-37}$$

where R_{1n} = curvature radius at critical point for equivalent normal helical pinion, in
R_{2n} = curvature radius at critical contact point for equivalent normal helical gear, in

The curvature radii are given by

$$R_{1n} = R_P - Z_c \tag{4-38}$$

$$\begin{aligned} R_{2n} &= R_G + Z \quad \text{external gears} \\ R_{2n} &= R_G - Z_c \quad \text{internal gears} \end{aligned} \tag{4-39}$$

where Eq. (4-38) applies to external gears and Eq. (4-39) to either, as appropriate. Also the term Z_c is obtained from Eq. (4-32).

The load-sharing ratio m_N is the ratio of the face width to the minimum total length of the contact lines:

$$m_N = \frac{F}{L_{\min}} \tag{4-40}$$

where m_N = load-sharing ratio
F = minimum net face width, in
$L_{\min}$ = minimum total length of contact lines, in

The calculation of $L_{\min}$ is a rather involved process. For most helical gears which have a face-contact ratio of at least 2.0, a conservative approximation for the load-sharing ratio ratio m_n may be obtained from

$$m_N = \frac{P_N}{0.95 Z} \tag{4-41}$$

where p_N = normal circular pitch in inches and Z = length of line of action in the transverse plane in inches. For helical gears with a face-contact ratio of less than 2.0,

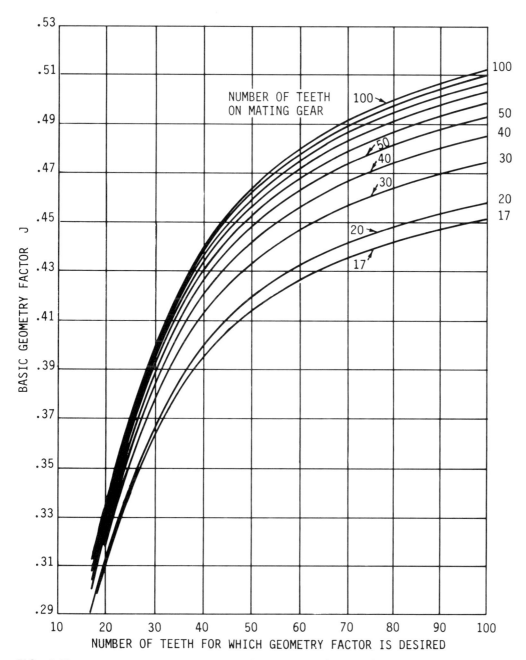

FIG. 4-11 Basic geometry factors for 20° spur teeth; $\phi_N = 20°$, $a = 1.00$, $b = 1.35$, $r_T = 0.42$, $\Delta t = 0$.

it is imperative that the actual value of L_{min} be calculated and used in Eq. (4-40). The method for doing this is shown in Eqs. (4-42) through (4-45):

$$L_{min} = \frac{1}{\sin \psi_b}[(P_1 - Q_1) + (P_2 - Q_2) + \cdots + (P_i - Q_i) + \cdots + (P_n - Q_n)] \quad (4\text{-}42)$$

where n = limiting number of lines of contact, as given by

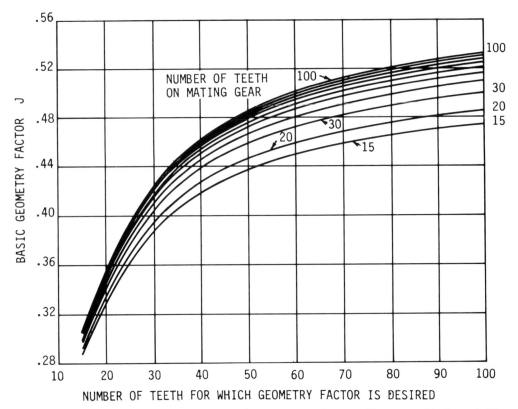

FIG. 4-12 Basic geometry factors for $22\frac{1}{2}°$ spur teeth; $\phi_N = 22\frac{1}{2}°$, $a = 1.00$, $b = 1.35$, $r_T = 0.34$, $\Delta t = 0$.

$$n = \frac{(Z/\tan \psi_b) + F}{p_x} \qquad (4\text{-}43)$$

Also P_i = sum of base pitches in inches. The ith term of P_i is the lesser of

$$ip_x \tan \psi_b \quad \text{or} \quad Z \qquad (4\text{-}44)$$

Finally, Q_i = remainder of base pitches in inches. Its value is

$$Q_i = 0 \quad \text{if} \quad ip_x \leq F$$

But when $ip_x > F$, then Q_i is the ith term and is the lesser of

$$(ip_x - F) \tan \psi_b \quad \text{or} \quad Z \qquad (4\text{-}45)$$

GEOMETRY FACTOR J. The bending strength geometry factor is

$$J = \frac{YC_\psi}{K_f m_N} \qquad (4\text{-}46)$$

where Y = tooth form factor
K_f = stress correction factor
C_ψ = helical factor
m_N = load-distribution factor

Notes · Drawings · Ideas

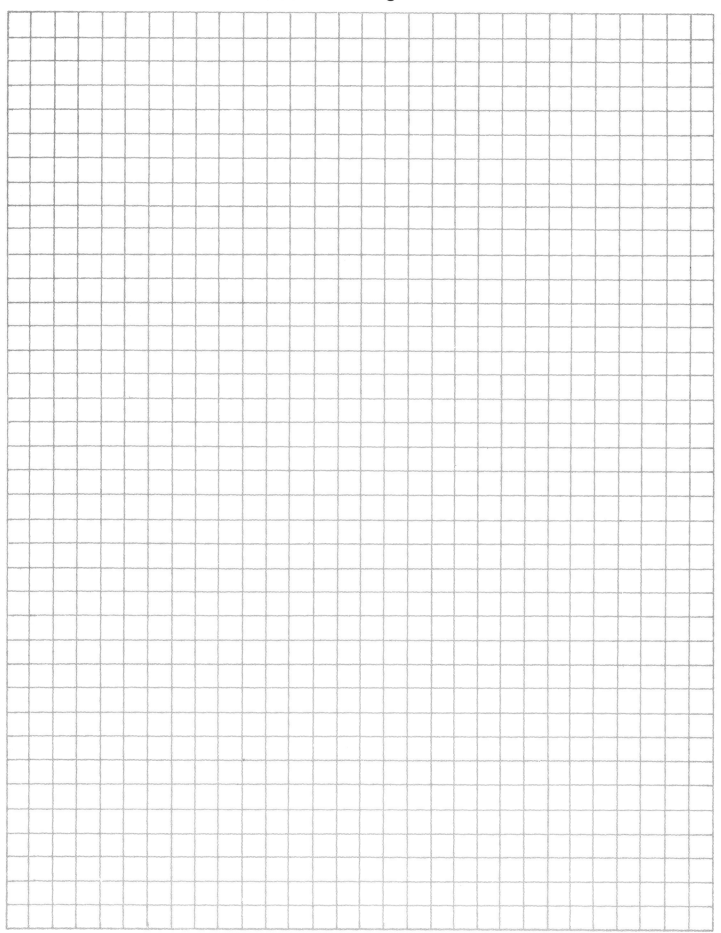

Both the helical and load-distribution factors were defined in the discussion of the geometry factor I. The calculation of Y is also a long, tedious process. For helical gears in which load sharing exists among the teeth in contact and for which the face-contact ratio is at least 2.0, the value of Y need not be calculated since the value for J may be obtained directly from the charts shown in Figs. 4-11 through 4-25 with Eq. (4-47):

$$J = J'Q_{TR}Q_{TT}Q_{A}Q_{H} \tag{4-47}$$

where J' = basic geometry factor
Q_{TR} = tool radius adjustment factor
Q_{TT} = tooth thickness adjustment factor
Q_{A} = addendum adjustment factor
Q_{H} = helix-angle adjustment factor

In using these charts, note that the values of addendum, dedendum, and tool-tip radius are given for a 1 normal pitch gear. Values for any other pitch may be obtained by dividing the factor by the actual normal diametral pitch. For example,

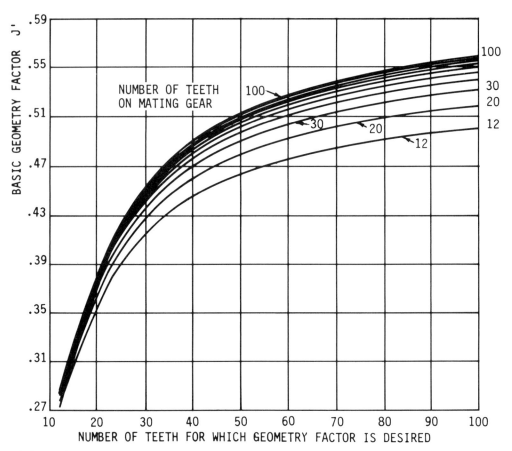

FIG. 4-13 Basic geometry factors for 25° spur teeth; $\phi_N = 25°$, $a = 1.00$, $b = 1.35$, $r_T = 0.24$, $\Delta t = 0$.

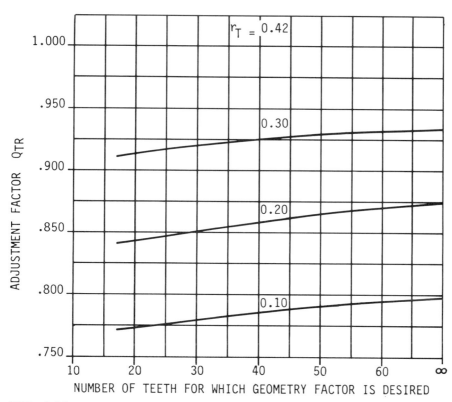

FIG. 4-14 Tool-tip radius adjustment factor for 20° spur teeth. Tool-tip radius = r_T for a 1 diametral pitch gear.

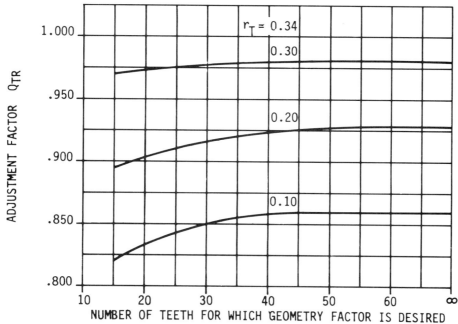

FIG. 4-15 Tool-tip radius adjustment factor for $22\frac{1}{2}°$ spur teeth. Tool-tip radius = r_T for a 1 diametral pitch gear.

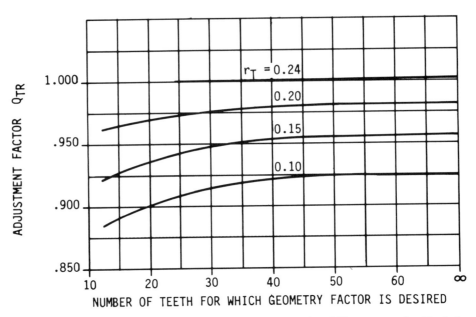

FIG. 4-16 Tool-tip radius adjustment factor for 25° spur teeth. Tool-tip radius = r_T for a 1 diametral pitch gear.

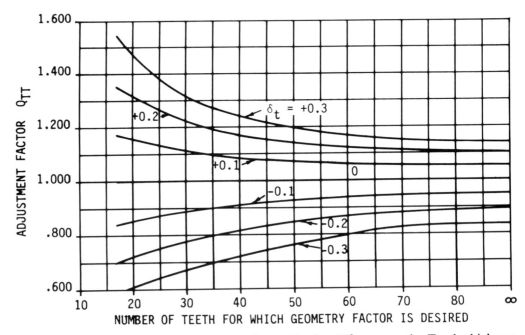

FIG. 4-17 Tooth thickness adjustment factor Q_{TT} for 20° spur teeth. Tooth thickness modification = δ_t for 1 diametral pitch gears.

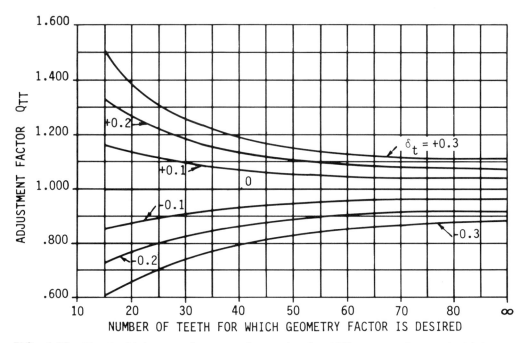

FIG. 4-18 Tooth thickness adjustment factor Q_{TT} for $22\frac{1}{2}°$ spur teeth. Tooth thickness modification = δ_t for 1 diametral pitch gears.

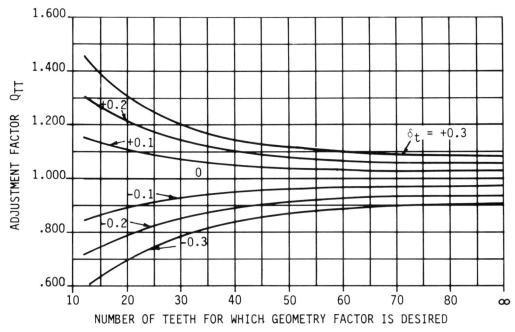

FIG. 4-19 Tooth thickness adjustment factor Q_{TT} for $25°$ spur teeth. Tooth thickness modification = δ_t for 1 diametral pitch gears.

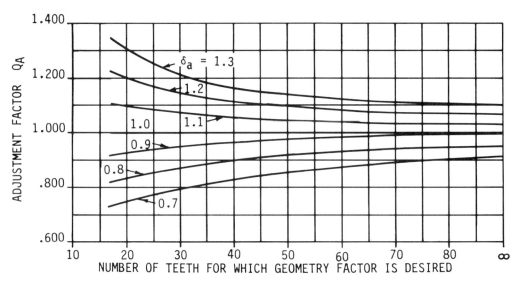

FIG. 4-20 Addendum adjustment factor Q_A for 20° spur teeth. Addendum factor modification = δ_a for 1 diametral pitch gears.

if an 8 normal pitch gear is being considered, the parameters shown on Fig. 4-11 are

$$\text{Addendum } a = \frac{1.0}{8} = 0.125 \text{ in}$$

$$\text{Dedendum } b = \frac{1.35}{8} = 0.168\,75 \text{ in}$$

$$\text{Tool (hob) tip radius } r_T = \frac{0.42}{8} = 0.0525 \text{ in}$$

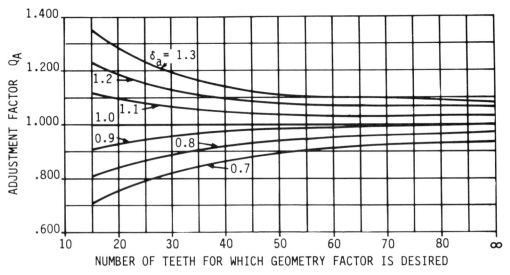

FIG. 4-21 Addendum adjustment factor Q_A for $22\frac{1}{2}°$ spur teeth. Addendum factor modification = δ_a for 1 diametral pitch gears.

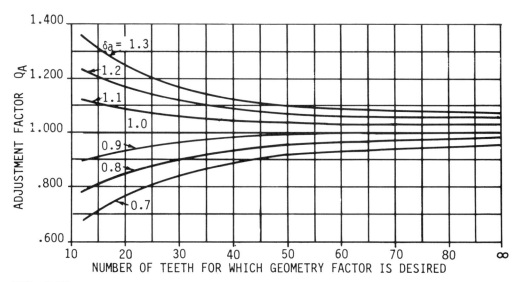

FIG. 4-22 Addendum adjustment factor Q_A for 25° spur teeth. Addendum factor modification = δ_a for 1 diametral pitch gears.

The basic geometry factor J' is found from Figs. 4-11 through 4-25. The tool radius adjustment factor Q_{TR} is found from Figs. 4-14 through 4-16 if the edge radius on the tool is other than $0.42/P_d$, which is the standard value used in calculating J'. Similarly, for gears with addenda other than $1.0/P_d$, or tooth thicknesses other than the standard value of $\pi/(2P_d)$, the appropriate factors may be obtained from these charts. In the case of a helical gear, the adjustment factor Q_H is obtained from Figs. 4-23 through 4-25. If a standard helical gear is being considered, Q_{TR}, Q_{TT}, and Q_A remain equal to unity, but Q_H must be found from Figs. 4-23 to 4-25.

These charts are computer-generated and, when properly used, produce quite accurate results. Also note that they are also valid for spur gears if Q_H is set equal to unity (that is, enter Figs. 4-23 through 4-25 with 0° helix angle).

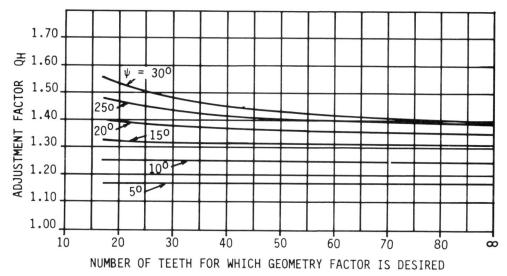

FIG. 4-23 Helix-angle adjustment factor Q_H for $\phi_N = 20°$.

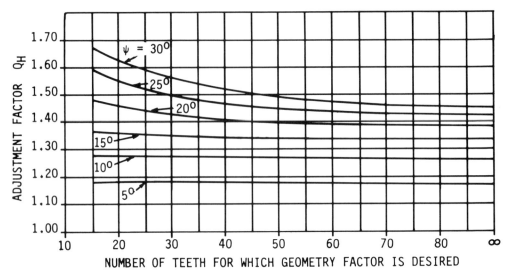

FIG. 4-24 Helix-angle adjustment factor Q_H for $\phi_N = 22\tfrac{1}{2}°$.

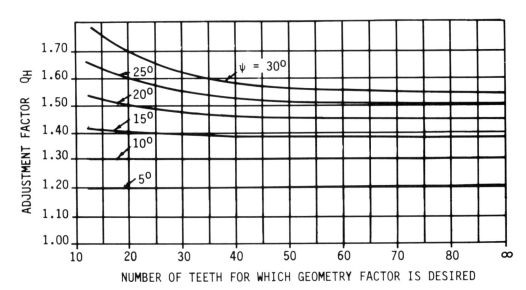

FIG. 4-25 Helix-angle adjustment factor Q_H for $\phi_N = 25°$.

The charts shown in Figs. 4-11 through 4-25 assume the use of a standard full-radius hob. Additional charts, still under the assumption that the face-contact ratio is at least 2.0 for other cutting tool configurations, are shown in Figs. 4-26 through 4-36.† For these figures

$$m_N = \frac{P_N}{0.95Z} \qquad (4\text{-}48)$$

where the value of Z is for an element of indicated number of teeth and a 75-tooth mate. Also, the normal tooth thicknesses of pinion and gear teeth are each reduced

---

†These figures are extracted from Ref. [4-1] with the permission of the AGMA.

Notes · Drawings · Ideas

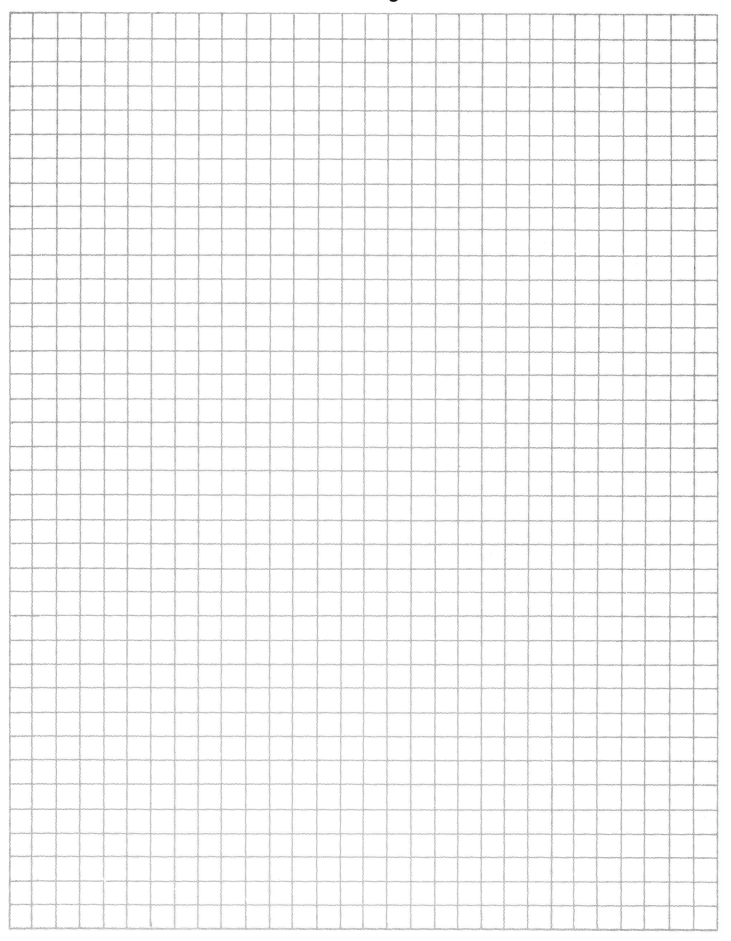

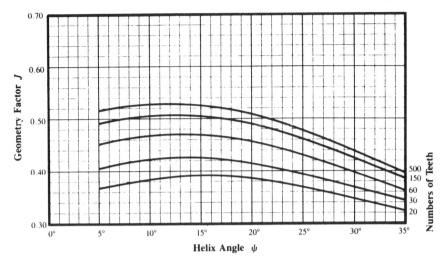

FIG. 4-26 Geometry factor J for a $14\frac{1}{2}°$ normal-pressure-angle helical gear. These factors are for a standard addendum finishing hob as the final machining operation. See Fig. 4-27a. *(From Ref. [4-1].)*

0.024 in, to provide 0.048 in of total backlash corresponding to a normal diametral pitch of unity. Note that these charts are limited to standard addendum, dedendum, and tooth thickness designs.

If the face-contact ratio is less than 2.0, the geometry factor must be calculated in accordance with Eq. (4-46); thus it will be necessary to define Y and K_f. The definition of Y may be accomplished either by graphical layout or by a numerical iteration procedure. Since this workbook is likely to be used by the machine designer with an occasional need for gear analysis, rather than the gear specialist, we present the direct graphical technique. Readers interested in preparing computer codes or calculator routines might wish to consult Ref. [4-6].

The following graphical procedure is abstracted directly from Ref. [4-1] with permission of the publisher, as noted earlier. The Y factor is calculated with the aid of dimensions obtained from an accurate layout of the tooth profile in the normal plane at a scale of 1 normal diametral pitch. Actually, any scale can be used, but the use of 1 normal diametral pitch is most convenient. Depending on the face-contact ratio, the load is considered to be applied at the highest point of single-tooth contact

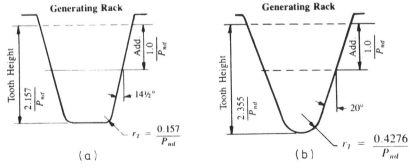

FIG. 4-27 Generating racks. (*a*) For teeth of Fig. 4-26; (*b*) for teeth of Fig. 4-29. *(From Ref. [4-1].)*

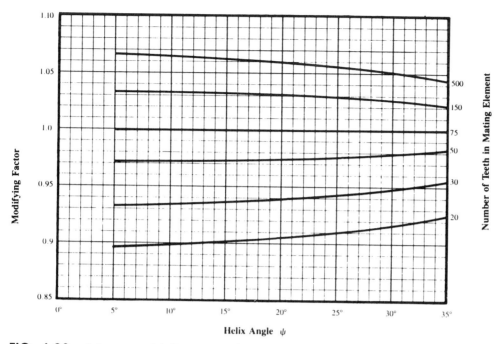

FIG. 4-28 J factor multipliers for $14\frac{1}{2}°$ normal-pressure-angle helical gear. These factors can be applied to the J factor when other than 75 teeth are used in the mating element. *(From Ref. [4-1].)*

(HPSTC), Fig. 4-37, or at the tooth tip, Fig. 4-37. The equation is

$$Y = \frac{K_\psi P_s}{[\cos(\phi_L)/\cos(\phi_{no})][(1.5/uC_h) - \tan(\phi_L)/t]} \qquad (4\text{-}49)$$

(see page 149 for definitions of terms).

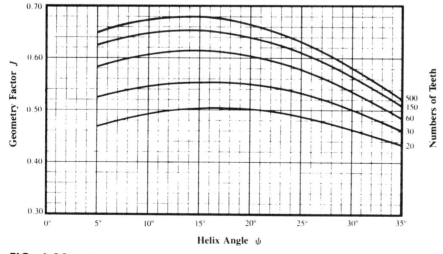

FIG. 4-29 Geometry factor J for a $20°$ normal-pressure-angle helical gear. These factors are for standard addendum teeth cut with a full-fillet hob. See Fig. 4-27*b*. *(From Ref. [4-1].)*

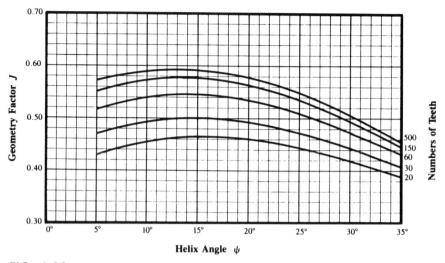

FIG. 4-30 Geometry factor J for a 20° normal-pressure-angle helical gear. These factors are for standard addendum teeth cut with a finishing hob as the final machining operation. See Fig. 4-31a. *(From Ref. [4-1].)*

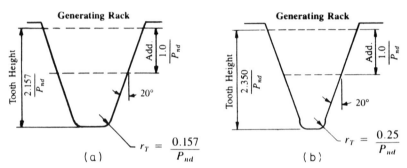

FIG. 4-31 Generating racks. (*a*) For teeth of Fig. 4-30; (*b*) for teeth of Fig. 4-32. *(From Ref. [4-1b].)*

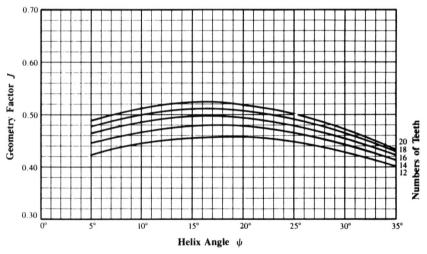

FIG. 4-32 Geometry factor J for 20° normal-pressure-angle helical gear. These factors are for long-addendum (125 percent of standard) shaved teeth cut with a preshave hob. See Fig. 4-31b. *(From Ref. [4-1].)*

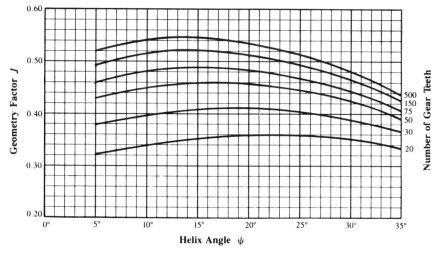

FIG. 4-33 Geometry factor J for 20° normal-pressure-angle helical gear. These factors are for short-addendum teeth (75 percent of standard) cut with a preshave hob. See Fig. 4-34. *(From Ref. [4-1].)*

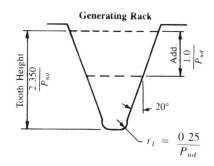

FIG. 4-34 Generating rack for teeth of Fig. 4-33. *(From Ref. [4-1].)*

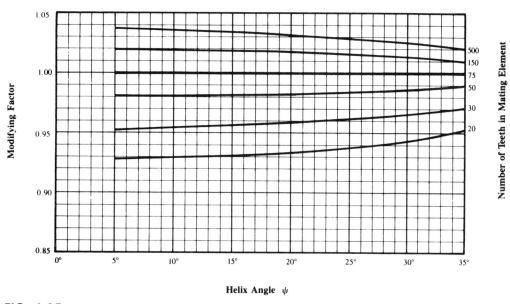

FIG. 4-35 J factor multipliers for 20° normal-pressure-angle helical gears. These factors can be applied to the J factor when other than 75 teeth are used in the mating element. *(From Ref. [4-1].)*

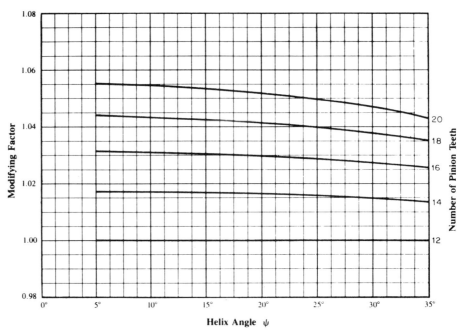

FIG. 4-36 J factor multipliers for 20° normal-pressure-angle helical gears with short addendum (75 percent of standard). These factors can be applied to the J factor when other than 75 teeth are used in the mating element. *(From Ref. [4-1].)*

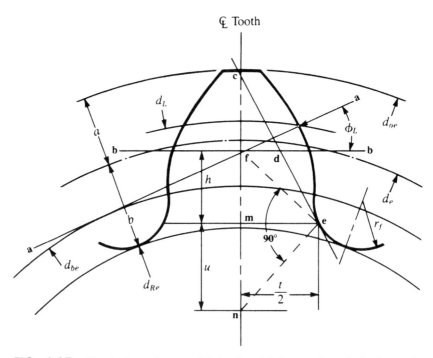

FIG. 4-37 Tooth form factor with load at highest point of single-tooth contact (HPSTC) shown in the normal plane through the pitch point. Note that r_f occurs at the point where the trochoid meets the root radius. *(From Ref. [4-1].)*

The terms in Eq. (4-49) are defined as follows:

- K_ψ = helix-angle factor
- ϕ_{No} = normal operating pressure angle [Eq. (4-14)]
- ϕ_L = load angle
- C_h = helical factor
- t = tooth thickness from layout, in
- u = radial distance from layout, in
- P_s = normal diametral pitch of layout (scale pitch), usually 1.0 in^{-1}

To make the Y factor layout for a helical gear, an equivalent normal-plane gear tooth must be created, as follows:

$$N_e = \frac{N_P}{\cos^3 \psi} \tag{4-50}$$

$$d_e = \frac{dP_{nd}}{\cos^2 \psi} \tag{4-51}$$

$$d_{be} = d_e \cos \phi_n = N_e \cos \phi_c \tag{4-52}$$

$$a = \frac{d_o - d}{2} P_{nd} \tag{4-53}$$

$$b = \frac{d - d_R}{2} P_{nd} \tag{4-54}$$

$$d_{oe} = d_e + 2a \tag{4-55}$$

$$d_{Re} = d_e - 2b \tag{4-56}$$

$$r_1 = \frac{(b - r_{Te})^2}{R_o + b - r_{Te}} \tag{4-57}$$

$$r_f = r_1 + r_{Te} \tag{4-58}$$

$$r_{Te} = r_T P_{nd} \tag{4-59}$$

For a hob or rack-shaped cutting tool;

$$R_o = \frac{d_{se}}{2} \tag{4-60}$$

For a pinion-shaped cutting tool;

$$R_o = \frac{d_{se} D_c P_{nd}}{2(d_{se} + D_c P_{nd})} \tag{4-61}$$

$$D_e = \frac{DP_{nd}}{\cos^2 \psi} \tag{4-62}$$

$$A = \frac{D_o - D}{2} P_{nd} \tag{4-63}$$

$$D_{oe} = D_e + 2A \tag{4-64}$$

$$D_{be} = D_e \cos \phi_n \tag{4-65}$$

$$d_{se} = \frac{N_e}{P_s} \tag{4-66}$$

where N_e = equivalent number of pinion teeth
d_{se} = equivalent generating pitch diameter, in
d_R = root diameter for actual number of teeth and generated pitch, in
d_{Re} = equivalent root diameter for equivalent number of teeth, in
d_{be} = equivalent base diameter for equivalent number of teeth, in
d_e = equivalent operating pitch diameter for equivalent number of teeth, in
d_{oe} = equivalent outside diameter for equivalent number of teeth, in
a = operating addendum of pinion at 1 normal diametral pitch, in
b = operating dedendum of pinion at 1 normal diametral pitch, in
r_f = minimum fillet radius at root circle of layout, in
r_T = edge radius of cutting tool, in
r_{Te} = equivalent edge radius of cutting tool, in
R_o = relative radius of curvature of pitch circle of pinion and pitch line or circle of cutting tool, in
D_c = pitch diameter of pinion-shaped cutting tool, in
D_e = equivalent operating pitch diameter of mating gear for equivalent number of teeth, in
D_{oe} = equivalent outside diameter of mating gear for equivalent number of teeth, in
D_{be} = equivalent base diameter of mating gear for equivalent number of teeth, in
A = operating addendum of mating gear at 1 normal diametral pitch, in

The dimensions defined by Eqs. (4-50) through (4-66) are then used to make a tooth-stress layout, as shown in either Fig. 4-37 or 4-38 as required by the face-contact ratio. That is, helical gears with low face-contact ratio ($m_F \leq 1.0$) are assumed to be loaded at the highest point of single-tooth contact; normal helical gears ($m_F > 1.0$) use tip loading, and the C_h factor compensates for the actual loading on the oblique line.

To find Y from the above data, a graphical construction, as follows, is required. For low-contact-ratio helical gears (with $m_F \leq 1.0$), using Fig. 4-37, draw a line $\overline{aa}$ through point p, the intersection of diameter d_L with the profile, and tangent to the base diameter d_{be}:

$$d_L = 2 \left\{ \left[\sqrt{\left(\frac{d_e}{2}\right)^2 - \left(\frac{d_{be}}{2}\right)^2} + Z_d \right]^2 + \left(\frac{d_{be}}{2}\right)^2 \right\}^{1/2} \qquad (4\text{-}67)$$

where Z_d = distance on line of action from highest point of single-tooth contact to pinion operating pitch circle, in inches, and so

$$Z_d = \pi \cos \phi_c - Z_e \qquad (4\text{-}68)$$

Letting Z_e = distance on line of action from gear outside diameter to pinion operating pitch circle, in inches, we have

$$Z_e = \sqrt{\left(\frac{D_{oe}}{2}\right)^2 - \left(\frac{D_{be}}{2}\right)^2} - \sqrt{\left(\frac{D_e}{2}\right)^2 - \left(\frac{D_{be}}{2}\right)^2} \qquad (4\text{-}69)$$

For normal helical gears with $m_F > 1.0$, using Fig. 4-38, we find $D_L = d_{oe}$. Draw a line $\overline{aa}$ through point p, the tip of the tooth profile, and tangent to the base diameter d_{be}. Continue the layout for all gear types as follows:

Through point f, draw a line $\overline{bb}$ perpendicular to the tooth centerline. The included angle between lines $\overline{aa}$ and $\overline{bb}$ is load angle ϕ_L.

Notes · Drawings · Ideas

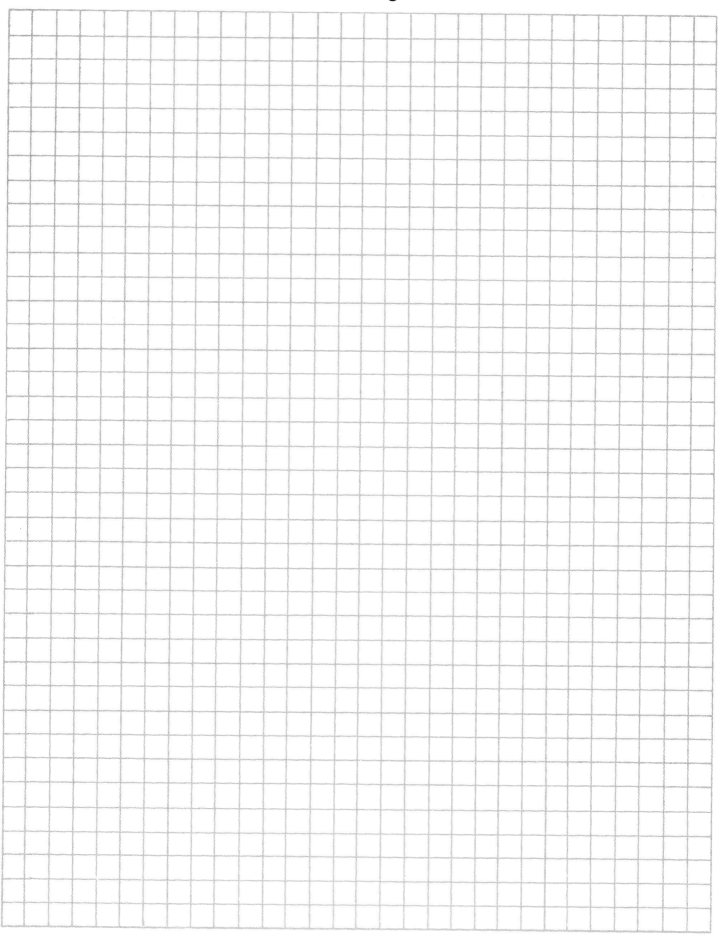

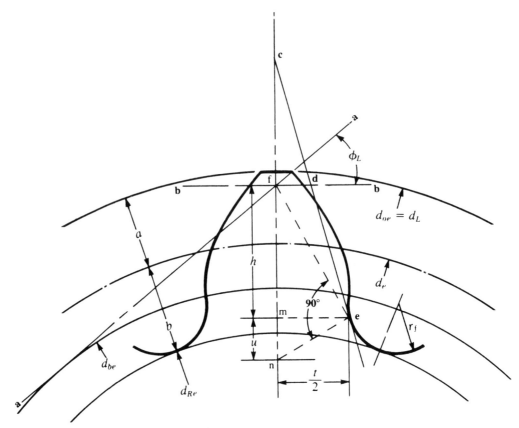

FIG. 4-38 Tooth form factor layout with load at tooth tip; shown in normal plane through the pitch point. *(From Ref. [4-1].)*

Draw line $\overline{cde}$ tangent to the tooth fillet radius r_f at e, intersecting line $\overline{bb}$ at d and the tooth centerline at c so that $\overline{cd} = \overline{de}$.

Draw line $\overline{fe}$.

Through point e, draw a line perpendicular to $\overline{fe}$, intersecting the tooth centerline at n.

Through point e, draw a line $\overline{me}$ perpendicular to the tooth centerline.

Measure the following in inches from the tooth layout:

$$\overline{mn} = u \qquad \overline{me} = \frac{t}{2}$$

and $\qquad \overline{mf} = h \qquad$ (required for calculating K_f)

The helix-angle factor K_ψ is set equal to unity for helical gears with $m_F \leq 1.0$, but for helical gears with $m_F > 1.0$ it is given by

$$K_\psi = \cos \psi_o \cos \psi \qquad (4\text{-}70)$$

where ψ_o = helix angle at operating pitch diameter [from Eq. (4-13)] and ψ = helix angle at standard pitch diameter.

The helical factor C_h is the ratio of the root bending moment produced by the same intensity of loading applied along the actual oblique contact line (Fig. 4-39). If the face width of one gear is substantially larger than that of its mate, then full

Notes • Drawings • Ideas

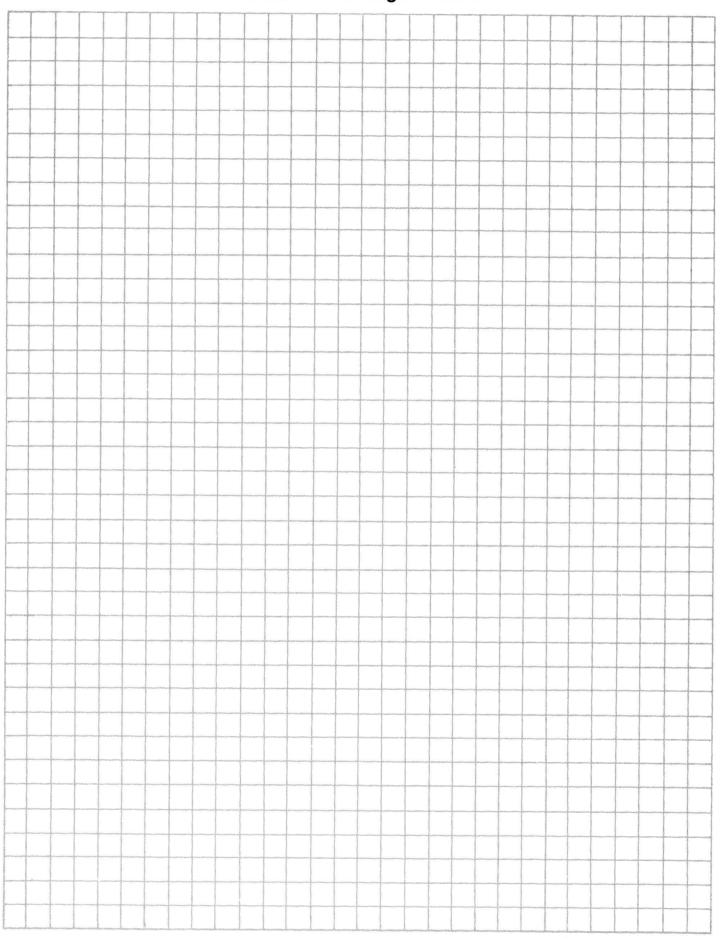

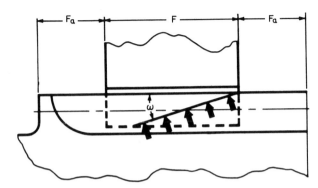

FIG. 4-39 Oblique contact line. Full buttressing exists when $F_a \geq$ one addendum.

buttressing may exist on the wider face gear. If one face is wider than its mate by at least one addendum on *both* sides, then the value of C_h defined below may be increased by 10 percent only. The helical factor is given by either Eq. (4-71) for low-contact-ratio helical gears ($m_F \leq 1.0$) or Eq. (4-72) for normal contact-ratio ($m_F > 1.0$) helicals. These equations are valid only for helix angles up to 30°:

$$C_h = 1.0 \tag{4-71}$$

$$C_h = \frac{1.0}{1 - [(\omega/100)(1 - \omega/100)]^{1/2}} \tag{4-72}$$

where $\omega = \tan^{-1}(\tan \psi_o \sin \phi_{No})$ = inclination angle, deg
ψ_o = helix angle at operating pitch diameter, deg [Eq. (4-13)]
ϕ_{No} = operating normal pressure angle, deg [Eq. (4-14)]

The tooth form factor Y may now be calculated from Eq. (4-49).

The stress correction factor is the last item which must be calculated prior to finding a value for the bending geometry factor J. Based on photoelastic studies by Dolan and Broghamer, the empirical relations shown in Eqs. (4-73) through (4-76) were developed:

$$K_f = H + \left(\frac{t}{r_f}\right)^L \left(\frac{t}{h}\right)^m \tag{4-73}$$

$$H = 0.18 - 0.008(\phi_{No} - 20) \tag{4-74}$$

$$L = H - 0.03 \tag{4-75}$$

$$m = 0.45 + 0.010(\phi_{No} - 20) \tag{4-76}$$

ELASTIC COEFFICIENT C_p. This factor accounts for the elastic properties of various gear materials. It is given by Eq. (4-77). Table 4-4 provides values directly for C_p for various material combinations, for which Poisson's ratio is 0.30.

$$C_p = \left\{\frac{1}{\pi[(1 - \nu_P^2)/E_P + (1 - \nu_G^2)/E_G]}\right\}^{1/2} \tag{4-77}$$

where ν_P, ν_G = Poisson's ratio for pinion and gear, respectively
E_P, E_G = modulus of elasticity for pinion and gear, respectively

Notes · Drawings · Ideas

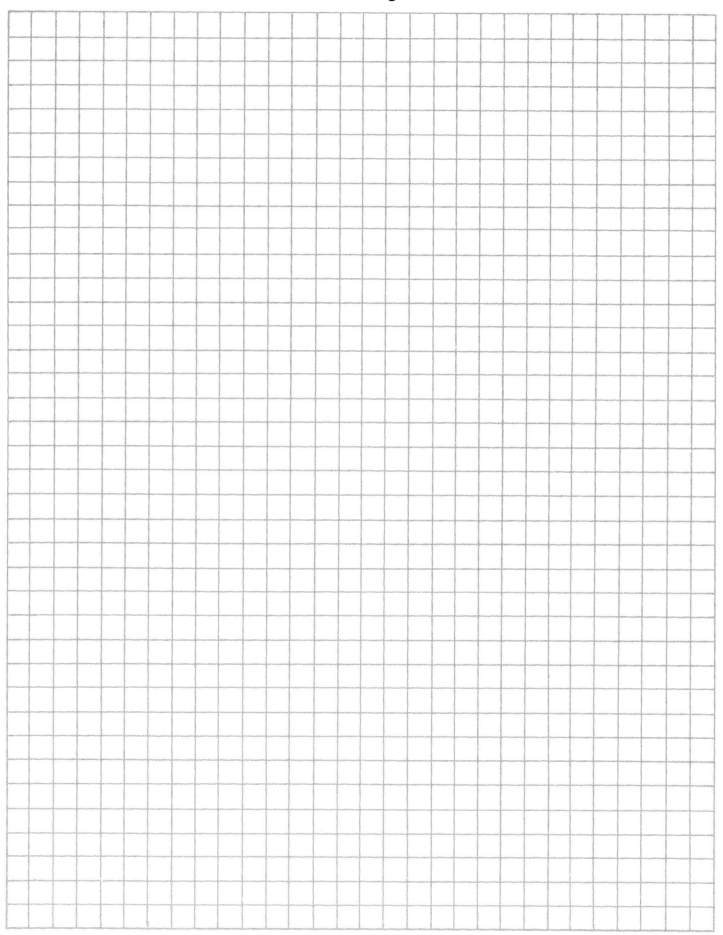

TABLE 4-4 Values of Elastic Coefficient C_p for Helical Gears with Nonlocalized Contact and for $v = 0.30$

Pinion material	Gear material					
	Steel	Malleable iron	Nodular iron	Cast iron	Aluminum bronze	Tin bronze
Steel, $E = 30$†	2300	2180	2160	2100	1950	1900
Malleable iron, $E = 25$	2180	2090	2070	2020	1900	1850
Nodular iron, $E = 24$	2160	2070	2050	2000	1880	1830
Cast iron, $E = 22$	2100	2020	2000	1960	1850	1800
Aluminum bronze, $E = 17.5$	1950	1900	1880	1850	1750	1700
Tin bronze, $E = 16$	1900	1850	1830	1800	1700	1650

†Modulus of elasticity E is in megapounds per square inch (Mpsi).

ALLOWABLE STRESSES s_{ac} AND s_{at}. The allowable stresses depend on many factors such as chemical composition, mechanical properties, residual stresses, hardness, heat treatment, and cleanliness. As a guide, the allowable stresses for helical gears may be obtained from Tables 4-5 and 4-6 or Figs. 4-40 and 4-41. Where a range of values is shown, the lowest values are used for general design. The upper values may be used only when the designer has certified that

1. High-quality material is used.
2. Section size and design allow maximum response to heat treatment.
3. Proper quality control is effected by adequate inspection.
4. Operating experience justifies their use.

Surface-hardened gear teeth require adequate case depth to resist the subsurface shear stresses developed by tooth contact loads and the tooth root fillet tensile stresses. But depths must not be so great as to result in brittle teeth tips and high residual tensile stress in the core.

The effective case depth for carburized and induction-hardened gears is defined as the depth below the surface at which the Rockwell C hardness has dropped to 50 R_C or to 5 points below the surface hardness, whichever is lower.

The values and ranges shown in Fig. 4-42 have had a long history of successful use for carburized gears and can be used as guides. For gearing in which maximum performance is required, detailed studies must be made of the application, loading, and manufacturing procedures, to obtain desirable gradients of both hardness and internal stress. Furthermore, the method of measuring the case, as well as the allowable tolerance in case depth, should be a matter of agreement between the customer and the manufacturer.

A guide for minimum effective case depth h_e at the pitch line for carburized and induction-hardened teeth, based on the depth of maximum shear from contact loading, is given by

$$h_e = \frac{C_G s_c d \sin \phi_o}{U_H \cos \psi_b} \qquad (4\text{-}78)$$

where h_e = minimum effective case depth in inches and U_H = hardening process factor in pounds per square inch. In Eq. (4-78), $U_H = 6.4 \times 10^6$ psi for carburized teeth and 4.4×10^6 psi for tooth-to-tooth induction-hardened teeth.

TABLE 4-5 Allowable Bending Stress Numbers s_{at} and Contact Stress Numbers s_{ac} for a Variety of Materials

AGMA class	Commercial designation	Heat treatment	Minimum hardness		s_{at}, kpsi	s_{ac}, kpsi
			Surface	Core		
Steel						
A-1 through A-5		Through-hardened and tempered (Fig. 35-40)	180 H_B and less		25–33	85–95
			240 H_B		31–41	105–115
			300 H_B		36–47	120–135
			360 H_B		40–52	145–160
			400 H_B		42–56	155–170
		Flame- or induction-hardened† with type A pattern (Fig. 35-45)	50–54 R_C		45–55	170–195
		Flame- or induction-hardened with type B pattern (Fig. 35-45)			22	
		Carburized† and case-hardened†	55 R_C		55–65	180–200
			60 R_C		55–70	200–225
	AISI 4140	Nitrided†‡	48 R_C	300 H_B	35–45	155–180
	AISI 4340	Nitrided†‡	46 R_C	300 H_B	36–47	150–175
	Nitralloy 135M	Nitrided†‡	60 R_C	300 H_B	38–48	170–195
	2½% chrome	Nitrided†‡	54–60 R_C	350 H_B	55–65	155–216
Cast iron						
20		As cast			5	50–60
30		As cast	175 H_B		8.5	65–75
40		As cast	200 H_B		13	75–85

TABLE 4-5 Allowable Bending Stress Numbers s_{at} and Contact Stress Numbers s_{ac} for a Variety of Materials (*Continued*)

AGMA class	Commercial designation	Heat treatment	Minimum hardness		s_{at}, kpsi	s_{ac}, kpsi
			Surface	Core		
Nodular (ductile) iron						
A-7-a A-7-c A-7-d A-7-e	60-40-18 80-55-06 100-70-03 120-90-02	Annealed, quenched, and tempered	140 H_B 180 H_B 230 H_B 270 H_B		0.90 to 1.00 times s_{at} for steel of same hardness	0.90 to 1.00 times s_{ac} for steel of same hardness
Malleable iron (pearlitic)						
A-8-c A-8-e A-8-f A-8-i	45007 50005 53007 80002		165 H_B 180 H_B 195 H_B 240 H_B		10 13 16 21	72 78 83 94
Bronze						
Bronze 2	AGMA 2C	Sand-cast	Min. tensile strength 40 kpsi		5.7	30
Al/Br 3	ASTM B-148-52 Alloy 9C	Heat-treated	Min. tensile strength 90 kpsi		23.6	65

†The range of allowable stress numbers indicated may be used with case depths as defined in the text.
‡The overload capacity of nitrided gears is low, since the shape of the effective SN curve is flat. The sensitivity to shock should be investigated before proceeding with the design.
SOURCE: Ref. [4-1].

TABLE 4-6 Reliability Factors K_R and C_R

Factor, K_R or C_R	Probabilities, %	
	Success	Failure
1.50	99.99	0.01
1.25	99.90	0.10
1.00	99.00	1.00
0.85	90.00	10.00

You should take care when using Eq. (4-78) that adequate case depths prevail at the tooth root fillet, and that tooth tips are not overhardened and brittle. A suggested value of maximum effective case depth $h_{e,\max}$ at the pitch line is

$$h_{e,\max} = \frac{0.4}{P_d} \quad \text{or} \quad h_{e,\max} = 0.56 t_o \tag{4-79}$$

where $h_{e,\max}$ = suggested maximum effective case depth in inches and t_o = normal tooth thickness at top land of gear in question, in inches.

For nitrided gears, case depth is specified as total case depth h_c, and h_c is defined as the depth below the surface at which the hardness has dropped to 110 percent of the core hardness.

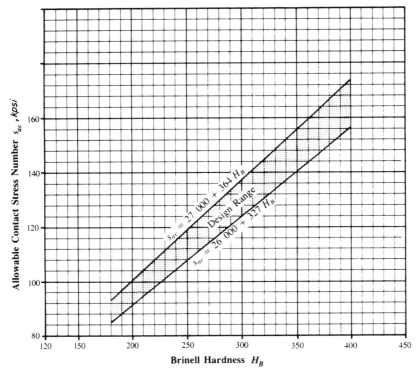

FIG. 4-40 Allowable contact stress number s_{ac} for steel gears. *(From Ref. [4-1].)*

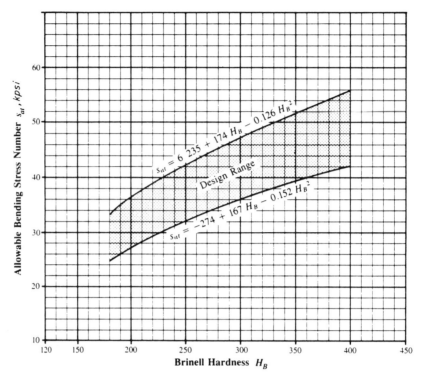

FIG. 4-41 Allowable bending stress number s_{at} for steel gears. *(From Ref. [4-1].)*

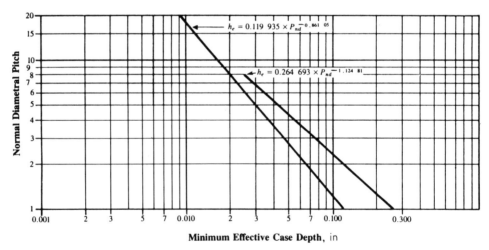

FIG. 4-42 Effective case depth h_e for carburized gears based on normal diametral pitch. The effective cash depth is defined as depth of case which has a minimum hardness of 50 R_C. The total case depth to core carbon is about $1.5 h_e$. The values and ranges shown on the case depth curves are to be used a guides. For gearing in which maximum performance is required, detailed studies must be made of the application, loading, and manufacturing procedures to obtain desirable gradients of both hardness and internal stress. Furthermore, the method of measuring the case as well as the allowable tolerance in case depth should be a matter of agreement between the customer and the manufacturer. *(From Ref. [4-1].)*

Notes · Drawings · Ideas

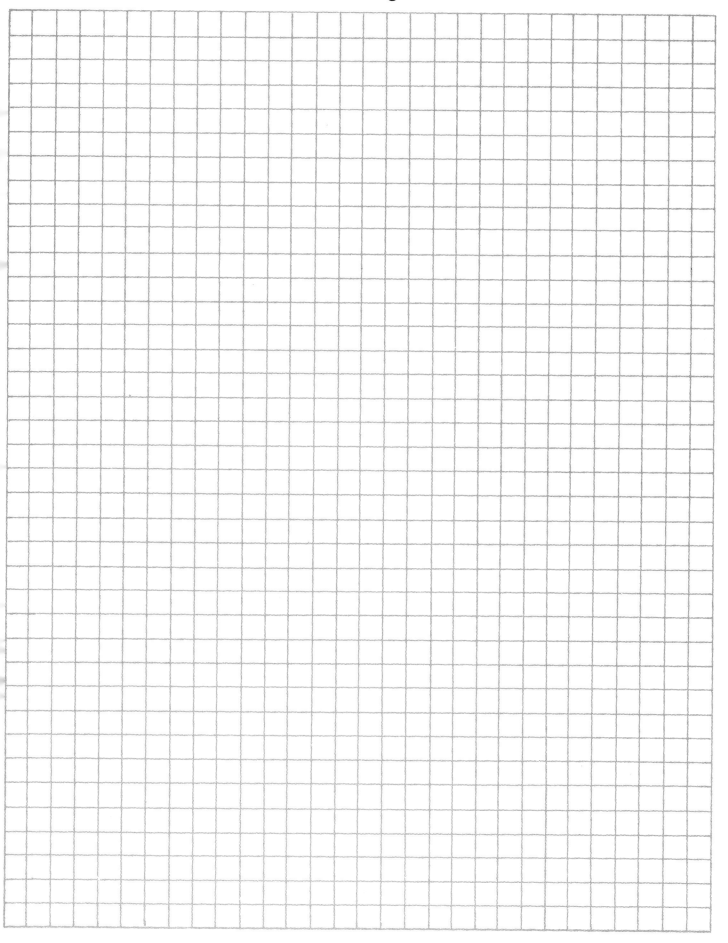

For gearing requiring maximum performance, especially large sizes, coarse pitches, and high-contact stresses, detailed studies must be made of application, loading, and manufacturing procedures to determine the desirable gradients of hardness, strength, and internal residual stresses throughout the tooth.

A guide for minimum case depth for nitrided teeth, based on the depth of maximum shear from contact loading, is given by

$$h_c = \frac{C_G U_c s_c d \sin \phi_o}{(1.66 \times 10^7)(\cos \psi_b)} \quad (4\text{-}80)$$

where h_c = minimum total case depth in inches and U_c = core hardness coefficient, from Fig. 4-43.

If the value of h_c from Eq. (4-80) is less than the value from Fig. 4-44, then the minimum value from Fig. 4-44 should be used. The equation for the lower or left-hand curve in Fig. 4-44 is

$$h_c = (4.328\ 96)(10^{-2}) - P_{nd}(9.681\ 15)(10^{-3}) + P_{nd}^2(1.201\ 85)(10^{-3})$$
$$- P_{nd}^3(6.797\ 21)(10^{-5}) + P_{nd}^4(1.371)(10^{-6}) \quad (4\text{-}81)$$

The equation of the right-hand curve is

$$h_c = (6.600\ 90)(10^{-2}) - P_{nd}(1.622\ 24)(10^{-2}) + P_{nd}^2(2.093\ 61)(10^{-3})$$
$$- P_{nd}^3(1.177\ 55)(10^{-4}) + P_{nd}^4(2.331\ 60)(10^{-6}) \quad (4\text{-}82)$$

Note that other treatments of the subject of allowable gear-tooth bending recommend that the value obtained from Table 4-6 or Fig. 4-41 be multiplied by 0.70 for teeth subjected to reversed bending. This is not necessary within the context of this analysis since the rim thickness factor K_b accounts for reversed bending.

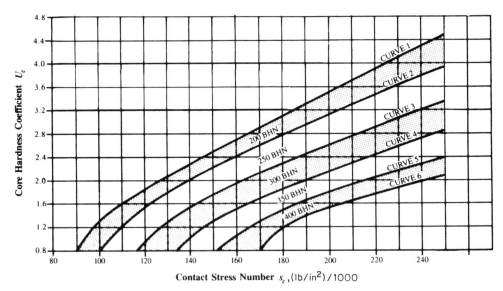

FIG. 4-43 Core-hardness coefficient U_c as a function of the contact stress number s_c. The upper portion of the core-hardness bands yields heavier case depths and is for general design purposes; use the lower portion of the bands for high-quality material. *(From Ref. [4-1].)*

Notes · Drawings · Ideas

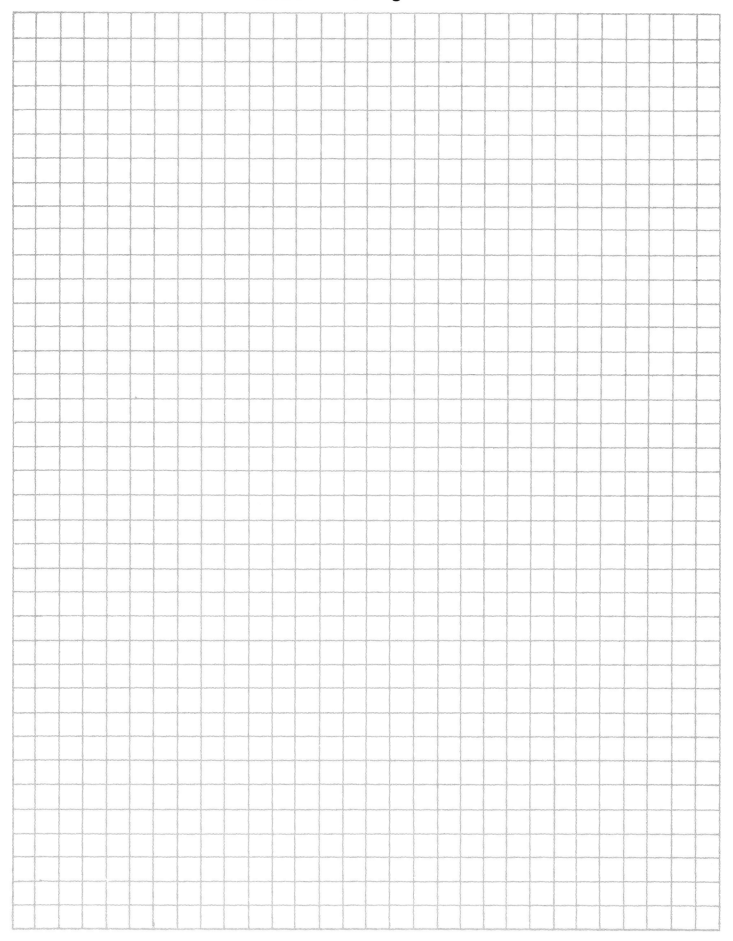

164 GEARING: A MECHANICAL DESIGNERS' WORKBOOK

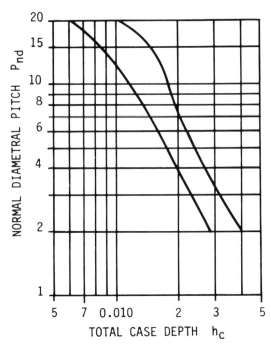

FIG. 4-44 Minimum total case depth h_c for nitrided gears based on the normal diametral pitch. *(From Ref. [4-1].)*

For through-hardened gears, the yield stress at maximum peak stress should also be checked as defined by Eq. (4-83):

$$S_{ay}K_y \geq \frac{W_{t,\max}K_a}{K_v}\frac{P_d}{F}\frac{K_m}{K_f} \qquad (4\text{-}83)$$

where $W_{t,\max}$ = peak tangential tooth load, lb
 K_a = application factor
 K_v = dynamic factor
 F = minimum net face width, in
 K_m = load-distribution factor
 K_f = stress correction factor
 K_y = yield strength factor
 s_{ay} = allowable yield strength number, psi (from Fig. 4-46)

The yield strength factor should be set equal to 0.50 for conservative practice or to 0.75 for general industrial use.

HARDNESS RATIO FACTOR C_H. It is common practice in using through-hardened gear sets to utilize a higher hardness on the pinion than on the gear. The pinion typically sees many more cycles than the gear; thus a more economical overall design is obtained by balancing the surface durability and wear rate in this manner. Similarly, surface-hardened pinions may be used with through-hardened gears to provide improved overall capacity through the work-hardening effect which a "hard" pinion has on a "soft" gear. The hardness ratio factor adjusts the allowable stresses for this effect.

HELICAL GEARS **165**

Spin Hardening

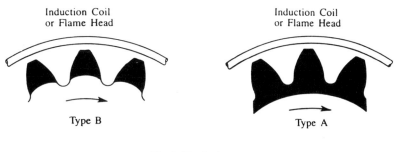

Flank Hardening

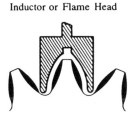
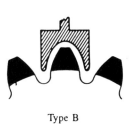

Flank and Root Hardening

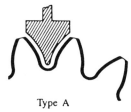

FIG. 4-45 Variations in hardening patterns obtainable with flame or induction hardening. *(From Ref. [4-1].)*

For through-hardened gear sets C_H can be found from Fig. 4-47 while Fig. 4-48 provides values for surface-hardened pinions mating with through-hardened gears.

LIFE FACTORS K_L AND C_L. The allowable stresses shown in Tables 4-5 and 4-6 and Figs. 4-40 and 4-41 are based on 10 000 000 load cycles. The life factor adjusts the allowable stresses for design lives other than 10 000 000 cycles. A unity value for the life factor may be used for design lives beyond 10 000 000 cycles only when it is justified by experience with similar designs.

Insufficient specific data are available to define life factors for most materials. For steel gears, however, experience has shown that the curves shown in Figs. 4-49 and 4-50 are valid.

In utilizing these charts, care should be exercised whenever the product of K_L and s_{ut} equals or exceeds s_{ay} as shown on Fig. 4-46, since this indicates that localized yielding may occur. For low-speed gears without critical noise vibration or transmission, accuracy requirements such as local yielding may be acceptable; but it should be avoided in general.

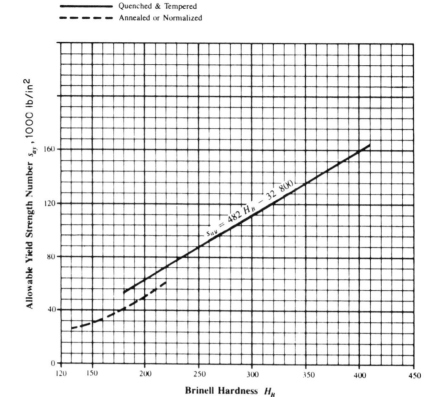

FIG. 4-46 Allowable yield strength number s_{ay} for steel gears. *(From Ref. [4-1].)*

RELIABILITY FACTORS C_R AND K_R. The allowable stress levels are not absolute parameters. Rather, a specific probability of failure is associated with each allowable level. The values shown in Figs. 4-40 and 4-41 and Table 4-5 are based on a 99 percent probability of success (or a 1 percent probability of failure). This means that in a *large* population at least 99 percent of the gears designed to a particular listed allowable stress will run for at least 10 000 000 cycles without experiencing a failure in the mode (that is, bending or durability) addressed.

In some cases it is desirable to design to higher or lower failure probabilities. Table 4-6 provides values for C_R and K_R which will permit the designer to do so. Before deciding on the reliability factor which is appropriate for a particular design, the analyst should consider what is meant by a "failure." In the case of a durability failure, a failure is said to have occurred when the first pit, or spall, is observed. Obviously a long time will elapse between the occurrence of a durability failure and the time at which the gear will cease to perform its normal power-transmission function. In the case of a bending failure, the appearance of a crack in the fillet area is the criterion. In most cases, and for most materials, the progression of this crack to the point at which a tooth or a piece of tooth fractures is rather quick. A bending failure will almost always progress to the point where function is lost much more rapidly than for a durability failure. For this reason sometimes it is desirable to use a higher value for K_R than for C_R.

Because of the load sharing which occurs on most normal helical gears, a complete fracture of a full single tooth, as often occurs on a spur gear, is not usually the mode

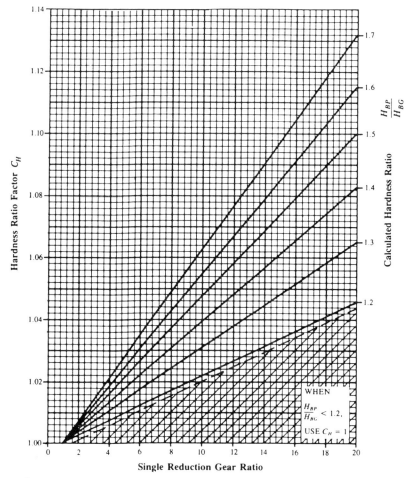

FIG. 4-47 Hardness ratio factor C_H for through-hardened gears. In this chart H_{BP} is the Brinell hardness of the pinion, and H_{BG} is the Brinell hardness of the gear. *(From Ref. [4-1].)*

of failure on a helical gear. A certain redundancy is built into a helical gear, since initially only a piece of a tooth will normally fracture.

TEMPERATURE FACTORS C_T AND K_T. At gear blank operating temperatures below 250°F and above freezing, actual operating temperature has little effect on the allowable stress level for steel gears; thus a temperature factor of unity is used. At higher or lower temperatures, the allowable stress levels are altered considerably. Unfortunately, few hard data are available to define these effects. At very low temperatures, the impact resistance and fracture toughness of most materials are reduced; thus special care must be exercised in such designs if nonuniform loading is expected. A temperature factor greater than unity should be used in such cases. Although no specific data are available, a value between 1.25 and 1.50 is recommended for gears which must transmit full power between 0 and −50°F.

At high temperatures, most materials experience a reduction in hardness level. Nonmetallic gears are not ordinarily used at high temperatures; thus our comments are restricted to steel gearing. The temperature factor should be chosen on the basis of the hot hardness curve for the particular material in use. That is, the temperature

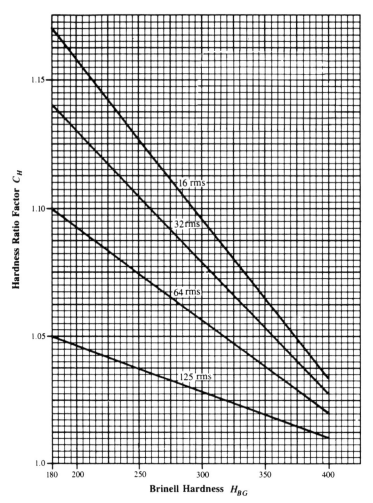

FIG. 4-48 Hardness ratio factor C_H for surface-hardened teeth. The rms values shown correspond to the surface finish of the pinion f_p in microinches. *(From Ref. [4-1].)*

factor is equal to the allowable stress at room-temperature hardness divided by the allowable stress at the hardness corresponding to the higher temperature. For information related to typical trends, Fig. 4-51 shows the hardness-temperature characteristics for two gear steels (AISI 9310 and VASCO-X2). Two typical bearing steels (M-50 and SAE 52100) are also shown for reference purposes.

Once the strength and durability analyses have been completed, the wear and scoring resistance of the gears must be defined. Wear is usually a concern only for relatively low-speed gears, while scoring is a concern only for relatively high-speed gears.

WEAR. Gear-tooth wear is a very difficult phenomenon to predict analytically. Fortunately, it is not a major problem for most gear drives operating in the moderate-to high-speed range. In the case of low-speed gears, however, not only is wear a significant problem, but also it can be the limiting factor in defining the load capacity of the mesh.

In low-speed gear drives, the film which separates the mating tooth surfaces is

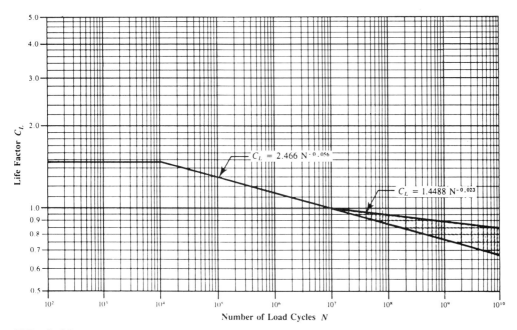

FIG. 4-49 Pitting-resistance life factor C_L. This curve does not apply where a service factor C_{SF} is used. *Note:* The choice of C_L above 10^7 cycles is influenced by lubrication regime, failure criteria, smoothness of operation required, pitch line velocity, gear material cleanliness, material ductility and fracture toughness, and residual stress. (*Source:* Ref. [4-1].)

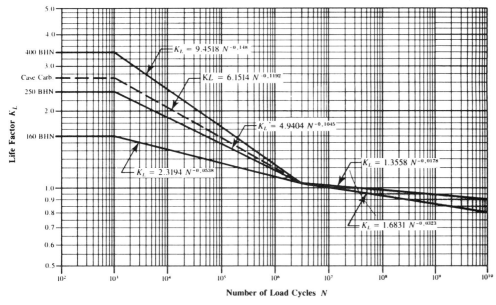

FIG. 4-50 Bending-strength life factor K_L. This chart does not apply where a service factor K_{SF} is used. *Note:* The choice of K_L above 3×10^6 cycles is influenced by pitch line velocity, gear material cleanliness, residual stress, gear material ductility, and fracture toughness. (*From Ref. [4-1].*)

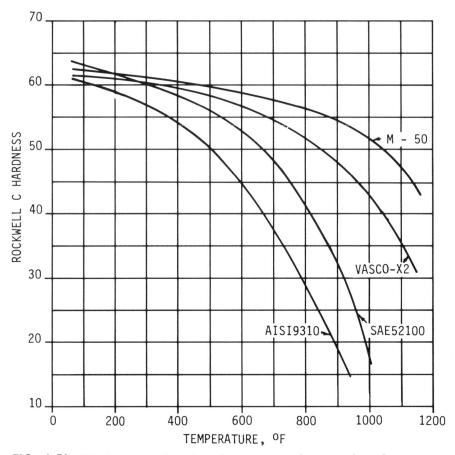

FIG. 4-51 Hardness as a function of temperature for several steels.

insufficient to prevent metal-to-metal contact; thus wear occurs. In higher-speed gears, the film becomes somewhat thicker, and gross contact of the mating surfaces is prevented. Indeed, grinding lines are still visible on many aircraft gears after hundreds of hours of operation. The type of surface distress which will occur in a gear set is dependent, to a certain extent, on the pitchline velocity. As shown in Fig. 4-52, wear predominates in the lower-speed range while scoring rules the upper-speed range. In the midrange, pitting controls the gear life.

The elastohydrodynamic (EHD) film thickness can provide some guidance in the evaluation of the wear potential of a gear set. Care must be used in the application of these methods, since the existing data are far from complete and there are many instances of contradictory results. One of the simplest approaches is due to Dowson; see Ref. [4-7]. The equation is

$$\frac{h}{R'} = \frac{(4.46 \times 10^{-5})(\alpha E')^{0.54}[\mu_o u/(E'R')]^{0.70}}{[w/(E'R')]^{0.13}} \quad (4\text{-}84)$$

where h = calculated minimum film thickness, in
R' = relative radius of curvature in transverse plane at pitch point, in
α = lubricant pressure-viscosity coefficient, in²/lb
E' = effective elastic modulus, psi

Notes · Drawings · Ideas

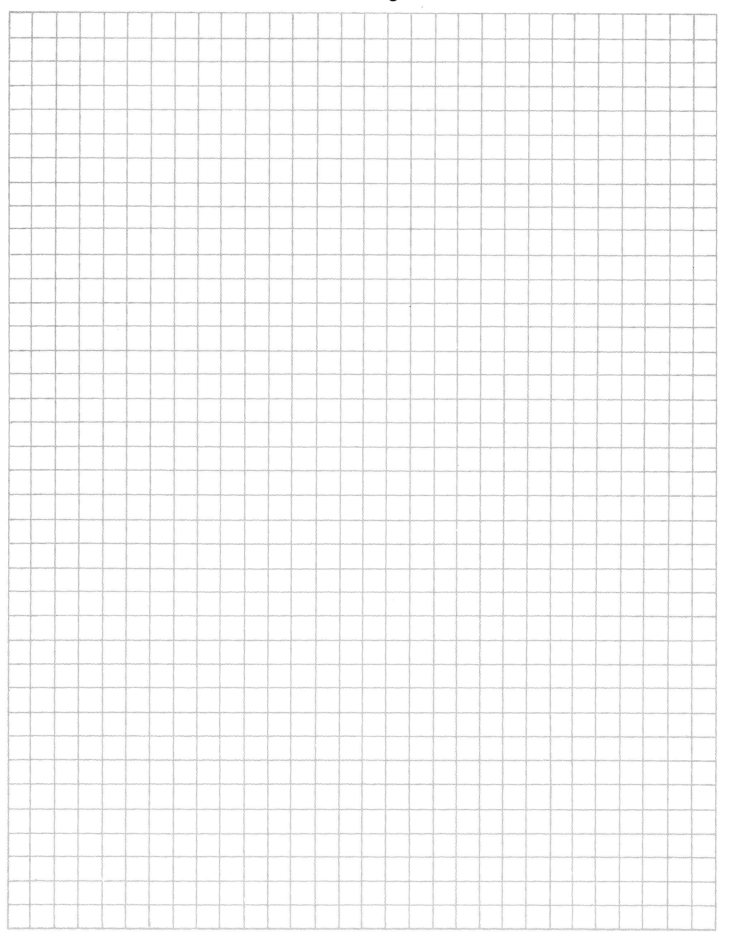

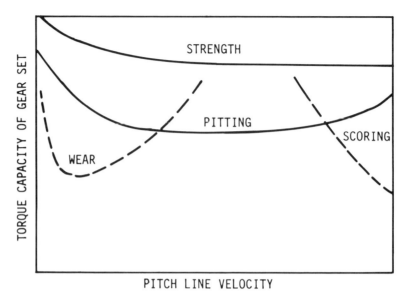

FIG. 4-52 Gear distress as a function of pitch line velocity.

μ_o = sump lubricant viscosity, centipoise (cP)
u = rolling velocity in transverse plane, inches per second (in/s)
w = load per unit length of contact, lb/in

Wellauer and Holloway ([4-8]) present a nomograph to compute the film thickness at the pitch point; but this nomograph is quite detailed and is not included here.

The parameter of interest in our discussion is not the film thickness itself, but rather the ratio of the film thickness to the relative surface roughness. This ratio is defined as the *specific film thickness* and is given by

$$\lambda = \frac{h}{S'} \tag{4-85}$$

The relative surface roughness, [root-mean-square (rms)] is given by

$$S' = \frac{S_P + S_G}{2} \tag{4-86}$$

Typical values for various gear manufacturing processes are shown in Table 4-7.

Once the specific film thickness has been determined, the probability of surface distress occurring can be determined through the use of Fig. 4-53.

Although the data presented thus far can be quite useful, several factors must be kept in mind in applying them to actual design. Most of the experimental data on which this information is based were obtained from through-hardened gear sets operating with petroleum-based oils. Gears operating with synthetic oils appear able to operate successfully at film thicknesses much less than those predicted by this analysis. The same is true for case-hardened gears of 59 R_C and higher hardness. The results may be further altered by the use of friction modifiers or EP additives in the oil. Finally, wear, of and by itself, is not necessarily a failure. In many cases, wear is an acceptable condition; it is simply monitored until it reaches some predetermined level, at which time the gears are replaced.

Perhaps the most useful application for this analysis is as a comparative, relative rating tool, rather than as an absolute design criterion.

TABLE 4-7 Tooth Surface Texture in the *As-Finished* Condition

Finish method	Surface texture in microinches (rms)	
	Range	Typical
Hobbed	30–80	50
Shaved	10–45	35
Lapped	20–200	93
Lapped and run in	20–100	53
Ground (soft)	5–35	25
Ground (hard)	5–35	15
Honed and polished	4–15	5

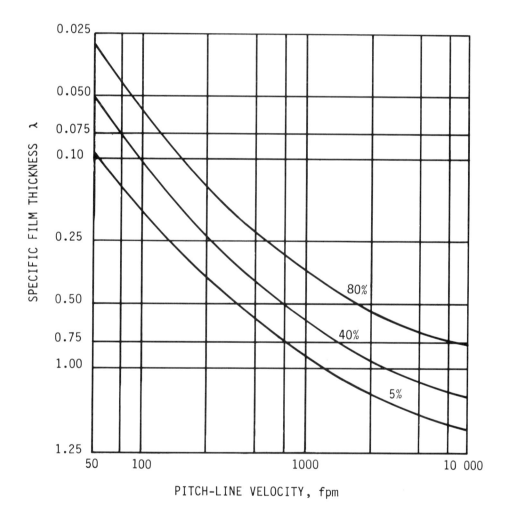

FIG. 4-53 Surface-distress probability chart as a function of pitch line velocity and specific film thickness. Curves represent 80, 40, and 5 percent probability of distress. The region above the 80 percent line is unsatisfactory; the region below the 5 percent line is good.

The occurrence of wear is difficult to predict, but the rate of wear is even more so. Equation (4-87) may be useful as a guide in predicting wear, but its accuracy has not been rigorously verified:

$$q = \frac{KW_t n_T}{FS_y} \tag{4-87}$$

where q = wear, in
 n_T = number of cycles
 S_y = yield strength of gear material, psi
 K = factor from Eq. (4-88)

and
$$3.1 \geq K\lambda^{1.645} \times 10^9 \geq 1.8 \tag{4-88}$$

In applying these equations, greater emphasis should be placed on the trend indicated than on the absolute value of the numbers. For example, a new design might be compared with an existing similar design for which the wear characteristics have been established. This could be accomplished by calculating the q value for each by Eqs. (4-87) and (4-88) and then comparing them, rather than looking at absolute values of either. The ratio of the two q values is far more accurate than the absolute value of either.

SCORING. Very few data are available in an easily usable form concerning the scoring behavior of gears. Scoring is normally a problem for heavily loaded, high-speed steel gears. The exact mechanism by which scoring occurs is not yet fully understood.

At high speeds, the calculated film thickness is often quite large. Yet a wearlike failure mode sometimes occurs. Under high-speed conditions, the sliding motion of one gear tooth on another may create instantaneous conditions of temperature and pressure which destroy the film of oil separating the tooth flanks. When this occurs, the asperities on the surface of the mating teeth instantaneously weld. As the gears continue to rotate, these welds break and drag along the tooth flanks, causing scratches, or "score" marks, in the direction of sliding. If the damage which occurs is very slight, it is often referred to as *scuffing,* or *frosting.* In some cases, light frosting may heal over and not progress; however, scoring is generally progressively destructive. Though never a catastrophic failure itself, scoring destroys the tooth surface, which leads to accelerated wear, pitting, and spalling. If scoring is allowed to progress unchecked, tooth fracture may ultimately occur.

Note that scoring is not a fatigue phenomenon; that is, its occurrence is not time-dependent. In general, if scoring does not occur within 15 to 25 minutes (min) at a certain operating condition, usually it will not occur *at that condition* at all. Only a change in operating condition, and not the accumulation of cycles, will cause scoring.

A theory known as the *critical-temperature theory,* originally proposed by Harmen Blok, is usually used in the evaluation of scoring hazard for a set of helical gears.

If we consider a simple analogy, the concept of critical temperature will become clear. Consider the old method of making fire by rubbing two sticks. If the sticks are held together with only light pressure and/or they are rubbed slowly, they will simply wear. If, however, the pressure is increased and the sticks are rubbed more rapidly, then the temperature at the mating surfaces will increase. If the pressure (load) and the rubbing speed (sliding velocity) are progressively increased, eventually the sticks will ignite. At the point of ignition, the sticks have reached their *critical* temperature. Quite obviously the critical temperature will vary with the type of wood, its moisture content, and other factors.

In a similar manner, as gear-tooth sliding velocity and load are increased, eventually a point will be reached at which the temperature at the conjunction attains a

Notes · Drawings · Ideas

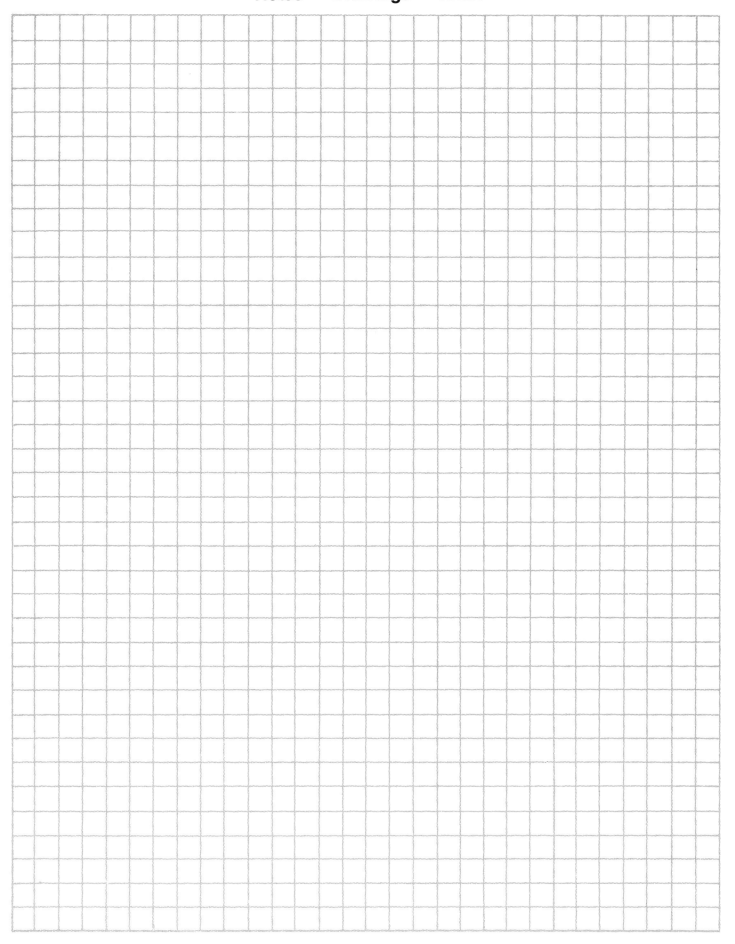

critical value, and then the film separating the tooth flanks will be destroyed. At this point the teeth are in metal-to-metal contact, and instantaneously welding of the surface asperities occurs. The continued rotation of the mesh rips apart these microscopic welds and produces the scored appearance from which this failure derives its name. The critical temperature varies with the type of gear material, surface hardness, surface finish, type and viscosity of oil, additives in the oil, etc. When the film is destroyed, it is sometimes referred to as *flashing;* thus the parameter used to evaluate this condition has come to be known as the *flash* temperature. When the flash temperature reaches its *critical* value, failure by scoring will occur. Note that the flash temperature referred to here is not related in any way to the flash point of the oil; and the oil flash point shown on some manufacturers' specification sheets is in no way related to the allowable flash temperature discussed here.

Many refinements have been made to Blok's original theory, and it is currently accepted as the best method available for evaluating scoring resistance for spur, helical, and bevel gears. Reference [4-9] presents a method of analysis for steel spur and helical gears based on Blok's method. The scoring hazard is evaluated by calculating a flash temperature rise ΔT_{Fi}. The flash temperature rise is added to the gear blank temperature T_B and compared with the allowable tooth flash temperature for the particular material and lubricant combination being used.

The flash temperature rise is given by

$$\Delta T_{Fi} = \left(\frac{W_{ti}C_aC_m}{FC_v}\right)^{0.75} \left(\frac{n_P^{0.5}}{P_d^{0.25}}\right) (\mu Z_{ti}) \left(\frac{50}{50 - S'}\right) \qquad (4\text{-}89)$$

where T_{Fi} = flash temperature rise at ith contact point along line of action, °F
W_{ti} = tangential tooth load at ith contact point, lb
F = net minimum face width, in
C_a = application factor
C_m = load-distribution factor
C_v = dynamic factor
n_P = pinion speed, revolutions per minute (rpm)
P_d = transverse diametral pitch
S' = relative surface roughness, Eq. (4-86)
Z_{ti} = scoring geometry factor at ith contact point along line of action

The factors C_a, C_v, and c_m are the same as those used in the durability formula [Eq. (4-17)].

The scoring geometry factor is given by

$$Z_{ti} = \frac{0.2917[\rho_{Pi}^{1/2} - (N_P\rho_{Gi}/N_G)^{1/2}]P_d^{1/4}}{(\cos \phi_i)^{0.75}[\rho_{Pi}\rho_{Gi}/(\rho_{Pi} + \rho_{Gi})]^{0.25}} \qquad (4\text{-}90)$$

where ρ_{Pi}, ρ_{Gi} = radius of curvature of pinion and gear, respectively, at ith contact point, in
N_P, N_G = tooth numbers of pinion and gear, respectively
P_d = transverse diametral pitch
ϕ_i = pressure angle at ith contact point, deg

The tooth flash temperature is then calculated by

$$T_{Fi} = T_B + \Delta T_{Fi} \qquad (4\text{-}91)$$

In most cases the blank temperature will be very close to the oil inlet temperature. Thus, unless the actual blank temperature is known, the oil inlet temperature may be used for T_B. Table 4-8 gives allowable values of the total flash temperature.

Notes · Drawings · Ideas

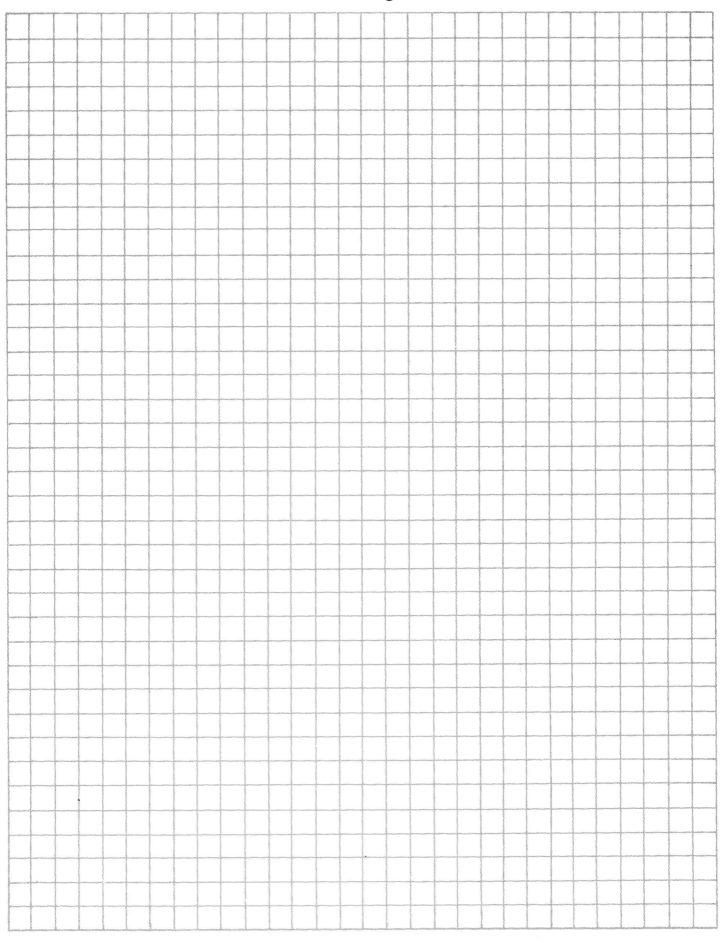

TABLE 4-8 Allowable Flash Temperatures for Some Gear Materials and for Spur and Helical Gears

The surface hardness is 60 R_C for all materials listed.

Gear material	Oil type	Allowable flash temperature, °F
AISI 9310	MIL-L-7808	295
	MIL-L-23699	295
	XAS 2354	335†
VASCO-X2	MIL-L-7808	350
	MIL-L-23699	350
	XAS 2354	375†

†Conservative estimate based on limited current data.

Equations (4-89) through (4-91) refer to the ith contact point. In utilizing these equations, the entire line of contact should be examined on a point-by-point basis to define the most critical contact point. Depending on the pitch of the tooth, 10 to 25 divisions should be adequate. For hand calculations this could be quite burdensome. A quick look at the highest and lowest points of single-tooth contact (based on a transverse-plane slice of the helical set) will provide a reasonable approximation.

The range of materials and oils shown in Table 4-8 is limited. Generally, scoring is a problem only in high-speed, high-load applications.

The most likely applications to be affected are aerospace types. This being the case, the material choice is limited to those shown, and usually either MIL-L-23699 or MIL-L-7808 oil is used. Some of the new XAS-2354 oils will provide much improved scoring resistance, but hard data are not presently available.

REFERENCES

4-1 AGMA, "Standard for Rating the Pitting Resistance and Bending Strength of Spur and Helical Involute Gear Teeth," American Gear Manufacturer's Association (AGMA), AGMA publ. 218.01, Dec. 1982.

4-2 "Design Guide for Vehicle Spur and Helical Gears," AGMA publ. 170.

4-3 *Gear Handbook*, vol. 1, *Gear Classification, Materials, and Measuring Methods for Unassembled Gears*, AGMA publ. 390.

4-4 Raymond J. Drago, "Results of an Experimental Program Utilized to Verify a New Gear Tooth Strength Analysis," AGMA publ. 229.27, Oct. 1983.

4-5 Raymond J. Drago, "An Improvement in the Conventional Analysis of Gear Tooth Bending Fatigue Strength," AGMA publ. 229.24, Oct. 1982.

4-6 R. Errichello, "An Efficient Algorithm for Obtaining the Gear Strength Geometry Factor on a Programmable Calculator," AGMA publ. 139.03, Oct. 1981.

4-7 D. Dowson, "Elastohydrodynamic Lubrication: Interdisciplinary Approach to the Lubrication of Concentrated Contacts," NASA SP-237, 1970.

4-8 E. J. Wellauer and G. Holloway, "Application of EHD Oil Film Theory to Industrial Gear Drives," ASME paper no. 75PTG-1, 1975.

4-9 "Informative Sheet—Gear Scoring Design Guide for Aerospace Spur and Helical Involute Gear Teeth," AGMA publ. 217.

chapter 5
WORM GEARING

K. S. EDWARDS, Ph.D.
Professor of Mechanical Engineering
University of Texas at El Paso
El Paso, Texas

GLOSSARY OF SYMBOLS

b_G Dedendum of gear teeth
C Center distance
d Worm pitch diameter
d_o Outside diameter of worm
d_R Root diameter of worm
D Pitch diameter of gear in central plane
D_b Base circle diameter
D_o Outside diameter of gear
D_t Throat diameter of gear
f Length of flat on outside diameter of worm
h_k Working depth of tooth
h_t Whole depth of tooth
L Lead of worm
m_G Gear ratio = N_G/N_W
m_o Module, millimeters of pitch diameter per tooth (SI use)
m_p Number of teeth in contact
n_w Rotational speed of worm, rpm
n_G Rotational speed of gear, rpm
N_G Number of teeth in gear
N_W Number of threads in worm
p_n Normal circular pitch
p_x Axial circular pitch of worm
P Transverse diametral pitch of gear, teeth per inch of diameter
W Force between worm and gear (various components are derived in the text)
λ Lead angle at center of worm, deg
ϕ_n Normal pressure angle, deg
ϕ_x Axial pressure angle, deg, at center of worm

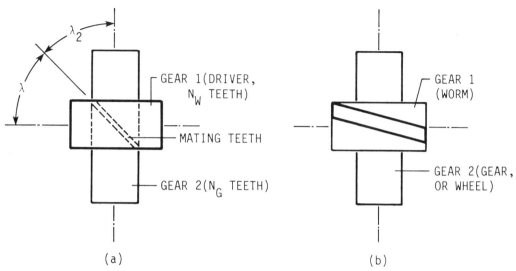

FIG. 5-1 (a) Helical gear pair; (b) a small lead angle causes gear one to become a worm.

5-1 INTRODUCTION

Worm gears are used for large speed reduction with concomitant increase in torque. They are limiting cases of helical gears, treated in Chap. 4. The shafts are normally perpendicular, though it is possible to accommodate other angles. Consider the helical-gear pair in Fig. 5-1a with shafts at 90°.

The lead angles of the two gears are described by λ (lead angle is 90° less the helix angle). Since the shafts are perpendicular, $\lambda_1 + \lambda_2 = 90°$. If the lead angle of gear 1 is made small enough, the teeth eventually wrap completely around it, giving the appearance of a screw, as seen in Fig. 5-1b. Evidently this was at some stage taken to resemble a *worm*, and the term has remained. The mating member is called simply the *gear*, sometimes the *wheel*. The helix angle of the gear is equal to the lead angle of the worm (for shafts at 90°).

The worm is always the driver in speed reducers, but occasionally the units are used in reverse fashion for speed increasing. Worm gear sets are self-locking when the gear cannot drive the worm. This occurs when the tangent of the lead angle is less than the coefficient of friction. The use of this feature in lieu of a brake is not recommended, since under running conditions a gear set may not be self-locking at lead angles as small as 2°.

There is only point contact between helical gears as described above. Line contact is obtained in worm gearing by making the gear envelop the worm as in Fig. 5-2; this is termed a *single-enveloping gear set*, and the worm is cylindrical. If the worm and gear envelop each other, the

FIG. 5-2 Photograph of a worm-gear speed reducer. Notice that the gear partially wraps, or envelopes, the worm. *(Cleveland Worm and Gear Company.)*

line contact increases as well as the torque that can be transmitted. The result is termed a *double-enveloping gear set*.

The minimum number of teeth in the gear and the reduction ratio determine the number of threads (teeth) for the worm. Generally, 1 to 10 threads are used. In special cases a larger number may be required.

5-2 KINEMATICS

In specifying the pitch of worm gear sets it is customary to state the axial pitch p_x of the worm. For 90° shafts this is equal to the transverse circular pitch of the gear. The advance per revolution of the worm, termed the lead L, is

$$L = p_x N_W$$

This and other useful relations result from consideration of the developed pitch cylinder of the worm, seen in Fig. 5-3. From the geometry, the following relations can be found:

$$d = \frac{N_W p_n}{\pi \sin \lambda} \tag{5-1}$$

$$d = \frac{N_W p_x}{\pi \tan \lambda} \tag{5-2}$$

$$\tan \lambda = \frac{L}{\pi d} = \frac{N_W p_x}{\pi d} \tag{5-3}$$

$$p_x = \frac{p_n}{\cos \lambda} \tag{5-4}$$

$$D = \frac{p_x N_G}{\pi} = \frac{N_G p_n}{\pi \cos \lambda} \tag{5-5}$$

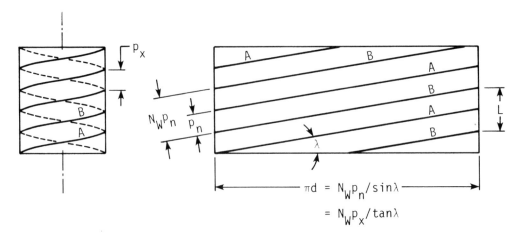

FIG. 5-3 Developed pitch cylinder of worm.

Notes · Drawings · Ideas

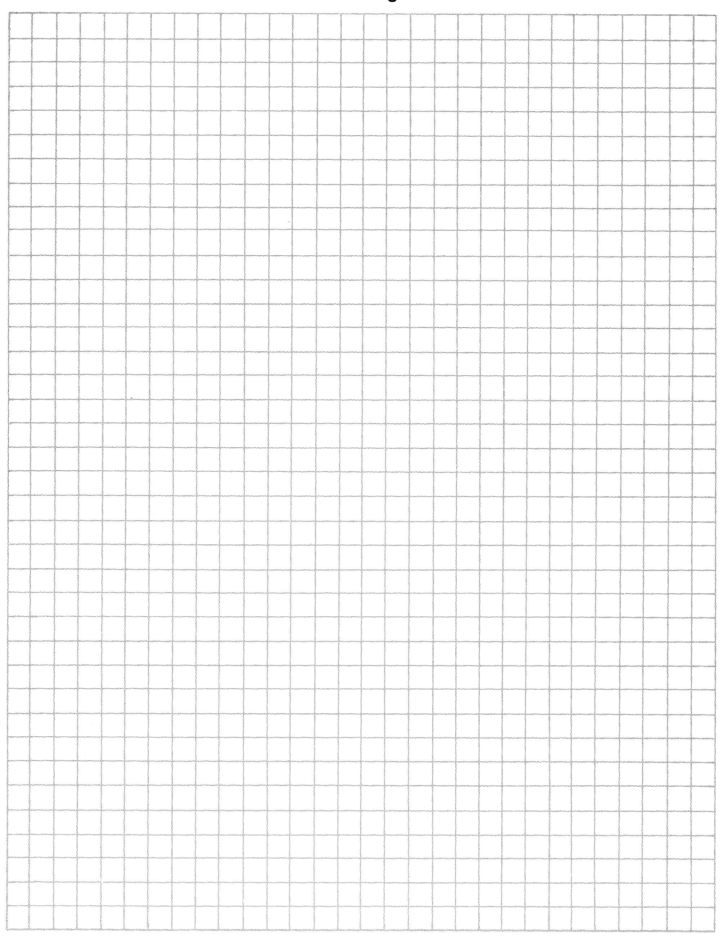

From Eqs. (5-1) and (5-5), we find

$$\tan \lambda = \frac{N_w d}{N_G D} = m_G \frac{d}{D} \qquad (5\text{-}6)$$

The center distance C can be derived from the diameters

$$C = \frac{p_n N_G}{2\pi}\left(\frac{m_G}{\cos \lambda} + \frac{1}{\sin \lambda}\right) \qquad (5\text{-}7)$$

which is sometimes more useful in the form

$$\frac{m_G}{\cos \lambda} + \frac{1}{\sin \lambda} = \begin{cases} \dfrac{2C}{p_n N_W} & \text{U.S. customary units} \\ \dfrac{2C}{m_o N_W} & \text{SI units} \\ \dfrac{2C}{d} & \text{either} \end{cases} \qquad (5\text{-}8)$$

For use in the International System (SI), recognize that

$$\text{Diameter} = N m_o = \frac{N p_x}{\pi}$$

so that the substitution

$$p_x = \pi m_o$$

will convert any of the equations above to SI units.

The pitch diameter of the gear is measured in the plane containing the worm axis and is, as for spur gears,

$$D = \frac{N_G p_x}{\pi} \qquad (5\text{-}9)$$

The worm pitch diameter is unrelated to the number of teeth. It should, however, be the same as that of the hob used to cut the worm-gear tooth.

5-3 VELOCITY AND FRICTION

Figure 5-4 shows the pitch line velocities of worm and gear. The coefficient of friction between the teeth μ is dependent on the sliding velocity. Representative values of μ are charted in Fig. 5-5. The friction has importance in computing the gear set efficiency, as will be shown.

5-4 FORCE ANALYSIS

If friction is neglected, then the only force exerted by the gear on the worm will be W, perpendicular to the mating tooth surface, shown in Fig. 5-6, and having the

Notes · Drawings · Ideas

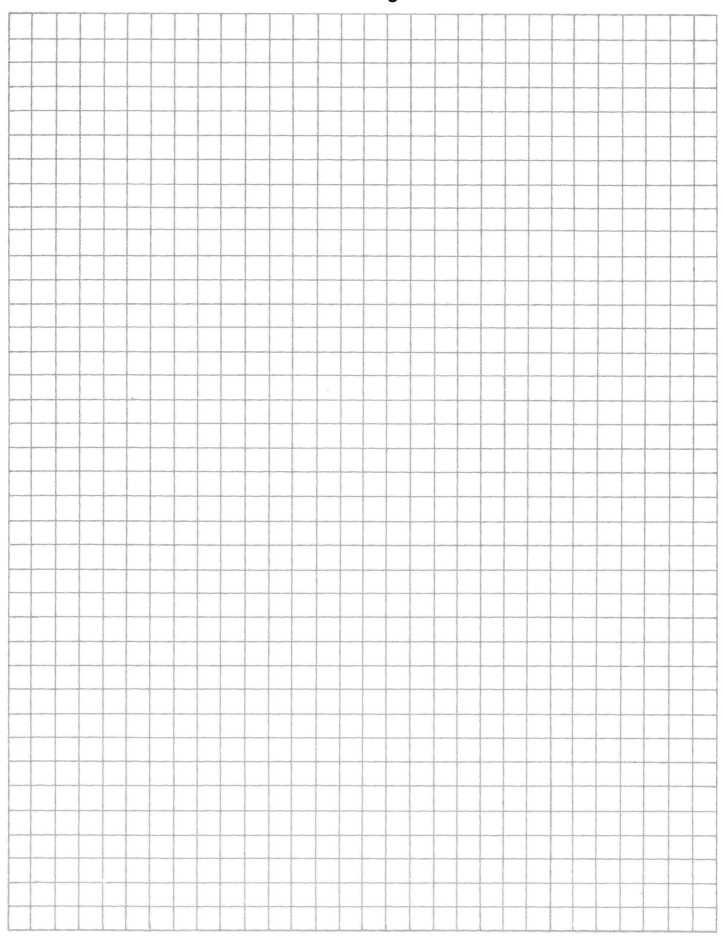

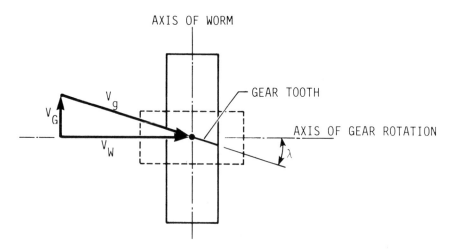

FIG. 5-4 Velocity components in a worm-gear set. The sliding velocity is
$V_S = (V_W^2 + V_G^2)^{1/2} = \dfrac{V_W}{\cos \lambda}$

three components W^x, W^y, and W^z. From the geometry of the figure

$$W^x = W \cos \phi_n \sin \lambda$$
$$W^y = W \sin \phi_n \qquad (5\text{-}10)$$
$$W^z = W \cos \phi_n \cos \lambda$$

In what follows, the subscripts W and G refer to forces *on* the worm and gear. The component W^y is the separating, or radial, force for both worm and gear (oppo-

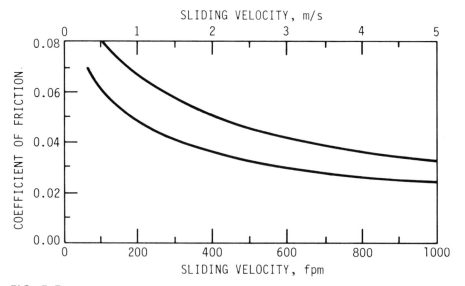

FIG. 5-5 Approximate coefficients of sliding friction between the worm and gear teeth as a function of the sliding velocity. All values are based on adequate lubrication. The lower curve represents the limit for the very best materials such as a hardened worm meshing with a bronze gear. Use the upper curve if moderate friction is expected.

Notes · Drawings · Ideas

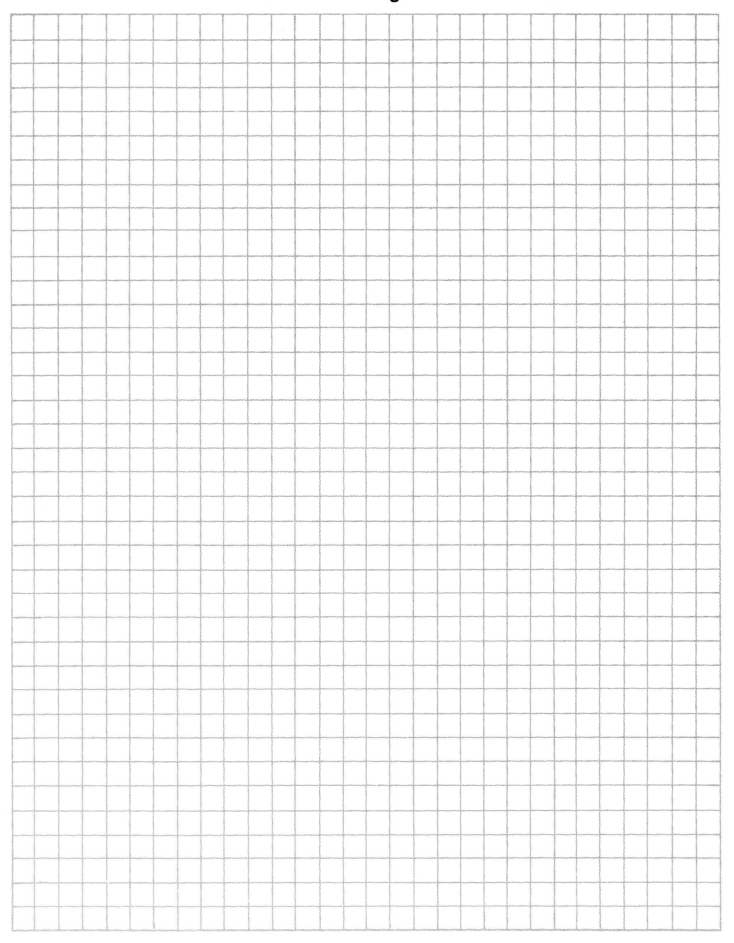

188 GEARING: A MECHANICAL DESIGNERS' WORKBOOK

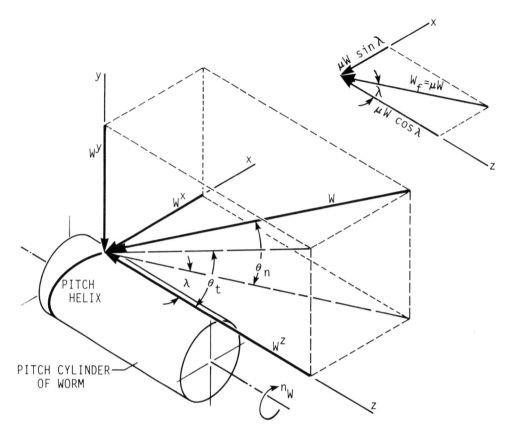

FIG. 5-6 Forces exerted on worm.

site in direction for the gear). The tangential force is W^x on the worm and W^z on the gear. The axial force is W^z on the worm and W^x on the gear. The gear forces are opposite to the worm forces

$$W_{W_t} = -W_{G_a} = W^x$$
$$W_{W_r} = -W_{G_r} = W^y \qquad (5\text{-}11)$$
$$W_{W_a} = -W_{G_t} = W^z$$

where the subscripts are t for the tangential direction, r for the radial direction, and a for the axial direction. It is worth noting in the above equations that the gear axis is parallel to the x axis and the worm axis is parallel to z. The coordinate system is right-handed.

The force W, which is normal to the profile of the mating teeth, produces a frictional force $W_f = \mu W$, shown in Fig. 5-6, along with its components $\mu W \cos \lambda$ in the negative x direction and $\mu W \sin \lambda$ in the positive z direction. Adding these to the force components developed in Eq. (5-10) yields

$$W^x = W(\cos \phi_n \sin \lambda + \mu \cos \lambda)$$
$$W^y = W \sin \phi_n \qquad (5\text{-}12)$$
$$W^z = W(\cos \phi_n \cos \lambda - \mu \sin \lambda)$$

Notes · Drawings · Ideas

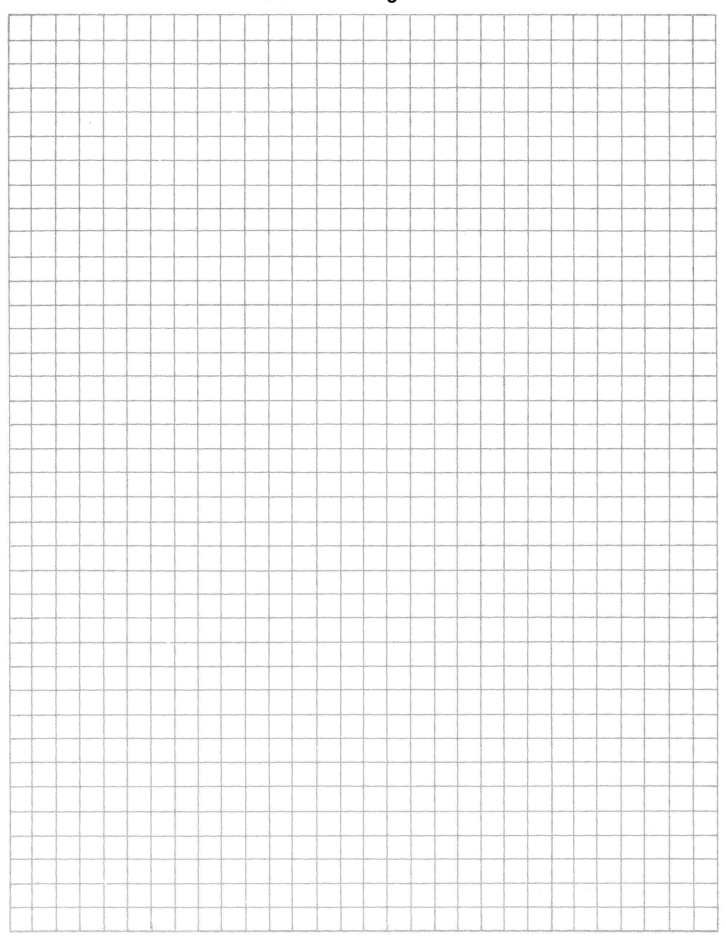

Equation (5-11) still applies. Substituting W^z from Eq. (5-12) into the third of Eq. (5-11) and multiplying by μ, we find the frictional force to be

$$W_f = \mu W = \frac{\mu W_{G_t}}{\mu \sin \lambda - \cos \phi_n \cos \lambda} \qquad (5\text{-}13)$$

A relation between the two tangential forces is obtained from the first and third of Eq. (5-11) with appropriate substitutions from Eq. (5-12):

$$W_{W_t} = W_{G_t} \frac{\cos \phi_n \sin \lambda + \mu \cos \lambda}{\mu \sin \lambda - \cos \phi_n \cos \lambda} \qquad (5\text{-}14)$$

The efficiency can be defined as

$$\eta = \frac{W_{W_t}(\text{without friction})}{W_{W_t}(\text{with friction})} \qquad (5\text{-}15)$$

Since the numerator of this equation is the same as Eq. (5-14) with $\mu = 0$, we have

$$\eta = \frac{\cos \phi_n - \mu \tan \lambda}{\cos \phi_n + \mu \cot \lambda} \qquad (5\text{-}16)$$

Table 5-1 shows how η varies with λ, based on a typical value of friction $\mu = 0.05$ and the pressure angles usually used for the ranges of λ indicated. It is clear that small λ should be avoided.

EXAMPLE 1. A 2-tooth right-hand worm transmits 1 horsepower (hp) at 1200 revolutions per minute (rpm) to a 30-tooth gear. The gear has a transverse diametral pitch of 6 teeth per inch. The worm has a pitch diameter of 2 inches (in). The normal pressure angle is $14\frac{1}{2}°$. The materials and workmanship correspond to the lower of the curves in Fig. 5-5. Required are the axial pitch, center distance, lead, lead angle, and the tooth forces.

Solution. The axial pitch is the same as the transverse circular pitch of the gear. Thus

$$p_x = \frac{\pi}{P} = \frac{\pi}{6} = 0.5236 \text{ in}$$

TABLE 5-1 Efficiency of Worm-Gear Sets for $\mu = 0.05$

Normal pressure angle ϕ_n, deg	Lead angle λ, deg	Efficiency η, percent
$14\frac{1}{2}$	1	25.2
	2.5	46.8
	5	62.6
	7.5	71.2
	10	76.8
	15	82.7
20	20	86.0
	25	88.0
	30	89.2

Notes • Drawings • Ideas

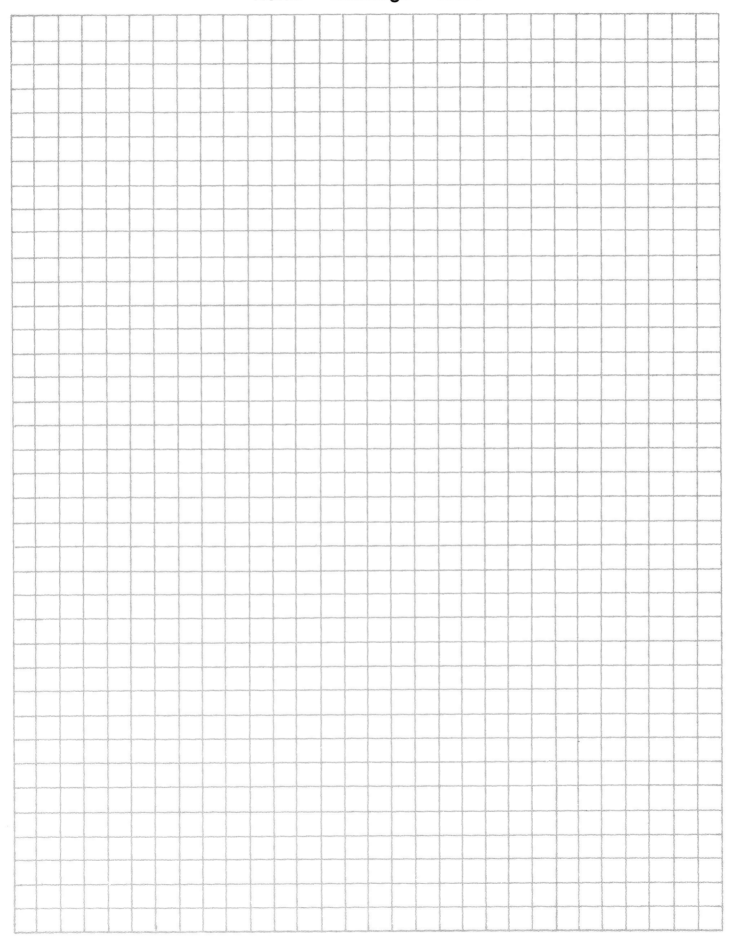

The pitch diameter of the gear is $D = N_G/P = 30/6 = 5$ in. The center distance is thus

$$C = \frac{D + d}{2} = \frac{2 + 5}{2} = 3.5 \text{ in}$$

The lead is

$$L = p_x N_W = 0.5236(2) = 1.0472 \text{ in}$$

From Eq. (5-3),

$$\lambda = \tan^{-1} \frac{L}{\pi d} = \tan^{-1} \frac{1.0472}{2\pi} = 9.46°$$

The pitch line velocity of the worm, in inches per minute, is

$$V_W = \pi d n_W = \pi(2)(1200) = 7540 \text{ in/min}$$

The speed of the gear is $n_G = 1200(2)/30 = 80$ rpm. The gear pitch line velocity is thus

$$V_G = \pi D n_G = \pi(5)(80) = 1257 \text{ in/min}$$

The sliding velocity is the square root of the sum of the squares of V_W and V_G, or

$$V_S = \frac{V_W}{\cos \lambda} = \frac{7540}{\cos 9.46} = 7644 \text{ in/min}$$

This result is the same as 637 feet per minute (fpm); we enter Fig. 5-5 and find $\mu = 0.03$.

Proceeding now to the force analysis, we use the horsepower formula to find

$$W_W = \frac{(33\,000)(12)(\text{hp})}{V_W} = \frac{(33\,000)(12)(1)}{7540} = 52.5 \text{ lb}$$

This force is the negative x direction. Using this value in the first of Eq. (5-12) gives

$$W = \frac{W^x}{\cos \phi_n \sin \lambda + \mu \cos \lambda}$$

$$= \frac{52.5}{\cos 14.5° \sin 9.46° + 0.03 \cos 9.46°} = 278 \text{ lb}$$

From Eq. (5-12) we find the other components of W to be

$$W^y = W \sin \phi_n = 278 \sin 14.5° = 69.6 \text{ lb}$$

$$W^z = W(\cos \phi_n \cos \lambda - \mu \sin \lambda)$$

$$= 278(\cos 14.5° \cos 9.46° - 0.03 \sin 9.46°)$$

$$= 265 \text{ lb}$$

The components acting on the gear become

$$W_{G_a} = -W^x = 52.5 \text{ lb}$$

$$W_{G_r} = -W^y = 69.6 \text{ lb}$$

$$W_{G_t} = -W^z = -265 \text{ lb}$$

The torque can be obtained by summing moments about the x axis. This gives, in inch-pounds,

$$T = 265(2.5) = 662.5 \text{ in·lb}$$

It is because of the frictional loss that this output torque is less than the product of the gear ratio and the input torque (778 lb·in).

5-5 STRENGTH AND POWER RATING

Because of the friction between the worm and gear, power is consumed by the gear set, causing the input and output horsepower to differ by that amount and resulting in a necessity to provide for heat dissipation from the unit. Thus

$$\text{hp(in)} = \text{hp(out)} + \text{hp(friction loss)}$$

This expression can be translated to the gear parameters, resulting in

$$\text{hp(in)} = \frac{W_{G_t} D n_W}{126\,000 m_G} + \frac{V_s W_f}{396\,000} \tag{5-17}$$

The force which can be transmitted W_{G_t} depends on tooth strength and is based on the gear, it being nearly always weaker than the worm (worm tooth strength can be computed by the methods used with screw threads, as in Chap. 1). Based on material strengths, an empirical relation is used. The equation is

$$W_{G_t} = K_s D^{0.8} F_e K_m K_v \tag{5-18}$$

where K_s = Materials and size correction factor, values for which are shown in Table 5-2

TABLE 5-2 Materials Factor K_s for Cylindrical Worm Gearing†

Face width of gear F_G, in	Sand-cast bronze	Static-chill-cast bronze	Centrifugal-cast bronze
Up to 3	700	800	1000
4	665	780	975
5	640	760	940
6	600	720	900
7	570	680	850
8	530	640	800
9	500	600	750

†For copper-tin and copper-tin-nickel bronze gears operating with steel worms case-hardened to 58 R_C minimum.

SOURCE: Darle W. Dudley (ed.), *Gear Handbook,* McGraw-Hill, New York, 1962, p. 13–38.

194 GEARING: A MECHANICAL DESIGNERS' WORKBOOK

TABLE 5-3 Ratio Correction Factor K_m

m_G	K_m	m_G	K_m	m_G	K_m
3.0	0.500	8.0	0.724	30.0	0.825
3.5	0.554	9.0	0.744	40.0	0.815
4.0	0.593	10.0	0.760	50.0	0.785
4.5	0.620	12.0	0.783	60.0	0.745
5.0	0.645	14.0	0.799	70.0	0.687
6.0	0.679	16.0	0.809	80.0	0.622
7.0	0.706	20.0	0.820	100.0	0.490

SOURCE: Darle W. Dudley (ed.), *Gear Handbook*, McGraw-Hill, New York, 1962, p. 13-38.

F_e = effective face width of gear; this is actual face width or two-thirds of worm pitch diameter, whichever is less
K_m = ratio correction factor, values in Table 5-3
Kv = velocity factor (Table 5-4)

EXAMPLE 2. A gear catalog lists a 4-pitch, $14\frac{1}{2}°$ pressure angle, single-thread hardened steel worm to mate with a 24-tooth sand-cast bronze gear. The gear has a $1\frac{1}{2}$-in face width. The worm has a 0.7854-in lead, 4.767° lead angle, $4\frac{1}{2}$-in face width, 3-in pitch diameter. Find the safe input horsepower.

TABLE 5-4 Velocity Factor K_v

Velocity V_S, fpm	K_v	Velocity V_S, fpm	K_v
1	0.649	600	0.340
1.5	0.647	700	0.310
10	0.644	800	0.289
20	0.638	900	0.269
30	0.631	1000	0.258
40	0.625	1200	0.235
60	0.613	1400	0.216
80	0.600	1600	0.200
100	0.588	1800	0.187
150	0.558	2000	0.175
200	0.528	2200	0.165
250	0.500	2400	0.156
300	0.472	2600	0.148
350	0.446	2800	0.140
400	0.421	3000	0.134
450	0.398	4000	0.106
500	0.378	5000	0.089
550	0.358	6000	0.079

SOURCE: Darle W. Dudley (ed.), *Gear Handbook*, McGraw-Hill, New York, 1962, p. 13-39.

From Table 5-2 $K_s = 700$. The pitch diameter of the gear is

$$D = \frac{N_G}{P} = \frac{24}{4} = 6 \text{ in}$$

The pitch diameter of the worm is given as 3 in; two-thirds of this is 2 in. Since the face width of the gear is smaller (1.5 in), $F_e = 1.5$ in. Since $m_G = N_G/N_W = 24/1 = 24$, from Table 5-3 $K_m = 0.823$ by interpolation. The pitch line velocity of the worm is

$$V_W = \pi d n_W = \pi(3)(1800) = 16\,965 \text{ in/min}$$

The sliding velocity is

$$V_S = \frac{V_W}{\cos \lambda} = \frac{16\,965}{\cos 4.767°} = 17\,024 \text{ in/min}$$

Therefore, from Table 5-4, $K_v = 0.215$. The transmitted load is obtained from Eq. (5-18) and is

$$W_{G_t} = K_s d^{0.8} F_e K_m K_v = 700(6^{0.8})(1.5)(0.823)(0.215)$$

$$= 779 \text{ lb}$$

To find the friction load, the coefficient of friction is needed. Converting V_S to feet per minute and using Fig. 5-5, we find $\mu = 0.023$. From Eq. (5-13) we find

$$W_f = \frac{\mu W_{G_t}}{\mu \sin \lambda - \cos \phi_n \cos \lambda}$$

$$= \frac{0.023(779)}{0.023 \sin 4.767° - \cos 14.5° \cos 4.767°}$$

$$= 18.6 \text{ lb}$$

Next, using Eq. (5-17), we find the input horsepower to be

$$\text{hp(in)} = \frac{W_{G_t} D n_W}{126\,000 m_G} + \frac{W_f V_S}{396\,000}$$

$$= \frac{779(6)(1800)}{126\,000(24)} + \frac{18.6(17\,024)}{396\,000}$$

$$= 2.78 + 0.80 = 3.58$$

5-6 HEAT DISSIPATION

In the last section we noted that the input and output horsepowers differ by the amount of power resulting from friction between the gear teeth. This difference represents energy input to the gear set unit, which will result in a temperature rise. The capacity of the gear reducer will thus be limited by its heat-dissipating capacity.

The cooling rate for rectangular housings can be estimated from

$$C_1 = \begin{cases} \dfrac{n}{84\,200} + 0.01 & \text{without fan} \\[2mm] \dfrac{n}{51\,600} + 0.01 & \text{with fan} \end{cases} \qquad (5\text{-}19)$$

where C_1 is the heat dissipated in Btu/(h)(in^2)(°F), British thermal units per hour–inch squared–degrees Fahrenheit and n is the speed of the worm shaft in rotations per minute. Note that the rates depend on whether there is a fan on the worm shaft. The rates are based on the area of the casing surface which can be estimated from

$$A_c = 43.2 C^{1.7} \qquad (5\text{-}20)$$

where A_c is in square inches.

The temperature rise can be computed by equating the friction horsepower to the heat-dissipation rate. Thus

$$\text{hp(friction)} = \frac{778 C_1 A_c \Delta T}{60(33\,000)} \qquad (5\text{-}21)$$

or

$$\Delta T(°F) = \frac{\text{hp(friction)}(60)(33\,000)}{778 C_1 A_c} \qquad (5\text{-}22)$$

The oil temperature should not exceed 180°F. Clearly the horsepower rating of a gear set may be limited by temperature rather than by gear strength. Both must be checked. Of course, other means than natural radiation and convection can be employed to solve the heat problem.

5-7 DESIGN STANDARDS

The American Gear Manufacturer's Association[†] has issued certain standards relating to worm-gear design. The purpose of these publications, which are the work of broad committees, is to share the experience of the industry and thus to arrive at good standard design practice. The following relate to industrial worm-gear design and are extracted from [5-1] with the permission of the publisher.

Gear sets with axial pitches of $\frac{3}{16}$ in and larger are termed *coarse-pitch*. Another standard deals with fine-pitch worm gearing, but we do not include these details here. It is not recommended that gear and worm be obtained from separate sources. Utilizing a worm design for which a comparable hob exists will reduce tooling costs.

[†]American Gear Manufacturer's Association (AGMA), 1500 King Street, Suite 201, Alexandria, Virginia 22314.

Notes · Drawings · Ideas

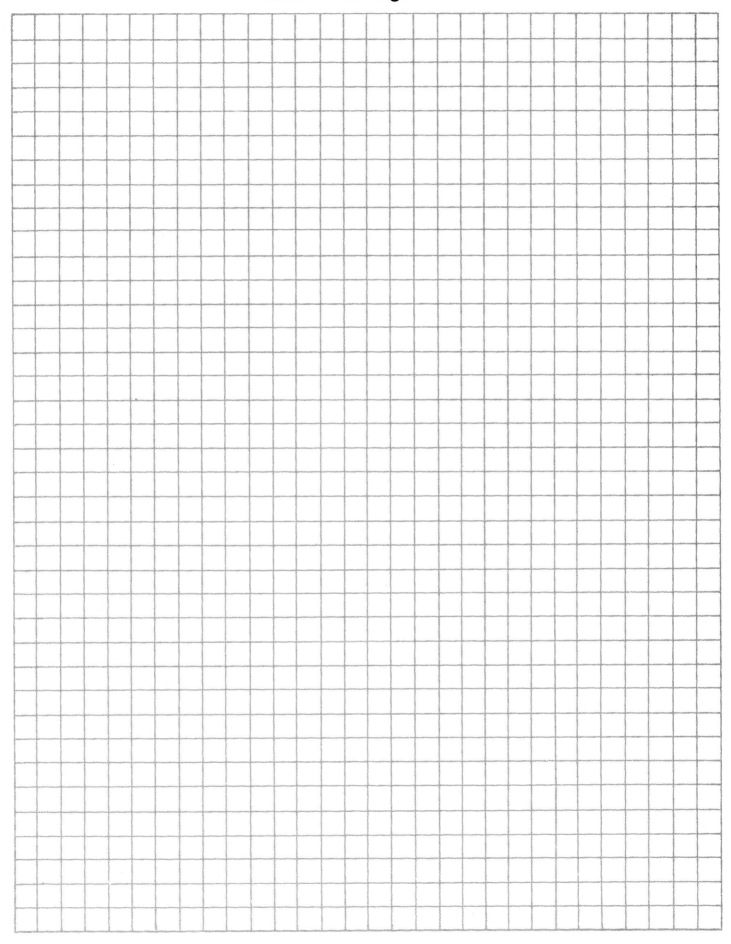

TABLE 5-5 Recommended Minimum Number of Gear Teeth

Center distance, in	Minimum number of teeth†
2	20
3	25
5	25
10	29
15	35
20	40
24	45

†Lower numbers are permissible for specific applications.

5-7-1 Number of Teeth of Gear

Center distance influences to a large extent the minimum number of teeth for the gear. Recommended minimums are shown in Table 5-5. The maximum number of teeth selected is governed by high ratios of reduction and considerations of strength and load-carrying capacity.

5-7-2 Number of Threads in Worm

The minimum number of teeth in the gear and the reduction ratio determine the number of threads for the worm. Generally, 1 to 10 threads are used. In special cases, a larger number may be required.

5-7-3 Gear Ratio

Either prime or even gear ratios may be used. However, if the gear teeth are to be generated by a single-tooth "fly cutter," the use of a prime ratio will eliminate the need for indexing the cutter.

5-7-4 Pitch

It is recommended that pitch be specified in the axial plane of the worm and that it be a simple fraction, to permit accurate factoring for change-gear ratios.

5-7-5 Worm Pitch Diameter

The pitch diameter of the worm for calculation purposes is assumed to be at the mean of the working depth. A worm does not have a true pitch diameter until it is mated with a gear at a specified center distance. If the actual addendum and dedendum of the worm are equal, respectively, to the addendum and dedendum of the gear, then the nominal and actual pitch diameters of the worm are the same. However, it is not essential that this condition exist for satisfactory operation of the gearing.

Although a relatively large variation in worm pitch diameter is permissible, it should be held within certain limits if the power capacity is not to be adversely affected. Therefore, when a worm pitch diameter is selected, the following factors should be considered:

1. Smaller pitch diameters provide higher efficiency and reduce the magnitude of tooth loading.
2. The root diameter which results from selection of a pitch diameter must be sufficiently large to prevent undue deflection and stress under load.
3. Larger worm pitch diameters permit utilization of larger gear face widths, providing higher strength for the gear set.
4. For low ratios the minimum pitch diameter is governed, to some degree, by the desirability of avoiding too high a lead angle. Generally, the lead is limited to a maximum of 45°. However, lead angles up to 50° are practical.

5-7-6 Gear Pitch Diameter

The selection of an approximate worm pitch diameter permits the determination of a corresponding approximate gear pitch diameter. In the normal case where the addendum and dedendum of the worm are to be equal, respectively, to the addendum and dedendum of the gear, a trial value of gear pitch diameter may be found by subtracting the approximate worm pitch diameter from twice the center distance of the worm and gear. Once the number of teeth for the gear has been selected, it is desirable to arrive at an exact gear pitch diameter by selecting for the gear circular pitch a fraction, which can be conveniently factored into a gear train for processing purposes, and calculating gear pitch diameter from the formula in Table 5-6. Should the actual value of gear pitch diameter differ from the trial value, the worm pitch diameter must be adjusted accordingly through the use of the formula in Table 5-7.

It is not essential that the pitch circle of the gear be at the mean of the working depth. Where there are sufficient teeth in the gear and the pressure angle is high enough to prevent undercutting, the pitch line can be anywhere between the mean of the depth and the throat diameter of the gear, or even outside the throat. This results in a short addendum for the gear teeth and lengthens the angle of recess. It is also practical for the gear pitch diameter to be located somewhat below the mean of the working depth.

TABLE 5-6 Dimensions of the Gear

Quantity	Symbol	Formula
Pitch diameter	D	$\dfrac{N_G p_x}{\pi}$
Throat diameter	D_t	$D + 2a$
Effective face width	F_e	$\sqrt{(d + h_k)^2 - d^2}$

TABLE 5-7 Dimensions of the Worm

Quantity	Symbol	Formula
Lead	l	$N_w P_x$
Pitch diameter†	d	$2C - D$
Outside diameter	d_o	$d + 2a$
Minimum face width	f	$2\sqrt{\left(\dfrac{D_t}{2}\right)^2 - \left(\dfrac{D}{2} - a\right)}$
Lead angle	λ	$\tan^{-1}\dfrac{l}{\pi d}$
Normal pitch	p_n	$p_x \cos \lambda$
Normal pressure angle	ϕ_n	See Table 5-6

†Use only where addenda and dedenda of worm and gear are equal.

5-7-7 Worm Thread and Gear-Tooth Proportions

PRESSURE ANGLE. Several factors deserve consideration in the selection of the pressure angle. Smaller values of pressure angle decrease the separating forces, extend the line of action, and result in less backlash change with change in center distance. Larger values of pressure angle provide stronger teeth and assist in preventing undercutting of the teeth where lead angles are larger. The recommended pressure angles are listed in Table 5-8. These, used with the system for stubbing teeth (Table 5-9) will avoid undercutting.

Although its use is discouraged, a 14° normal pressure angle may be used for lead angles up to 17°. A detailed study of gear-tooth action is employed by some designers to utilize pressure angles less than 25° where worm lead angles are above 30°.

TOOTH DEPTH PROPORTIONS. The choice of tooth depth proportions is governed, to a great extent, by the need to avoid undercutting of the gear teeth. Commonly used tooth depth proportions for lead angles to, but not including, 30° are

TABLE 5-8 Recommended Values for the Normal Pressure Angle

Normal pressure angle ϕ_n, deg	Lead angle λ, deg
20	Less than 30
25	30–45

Notes · Drawings · Ideas

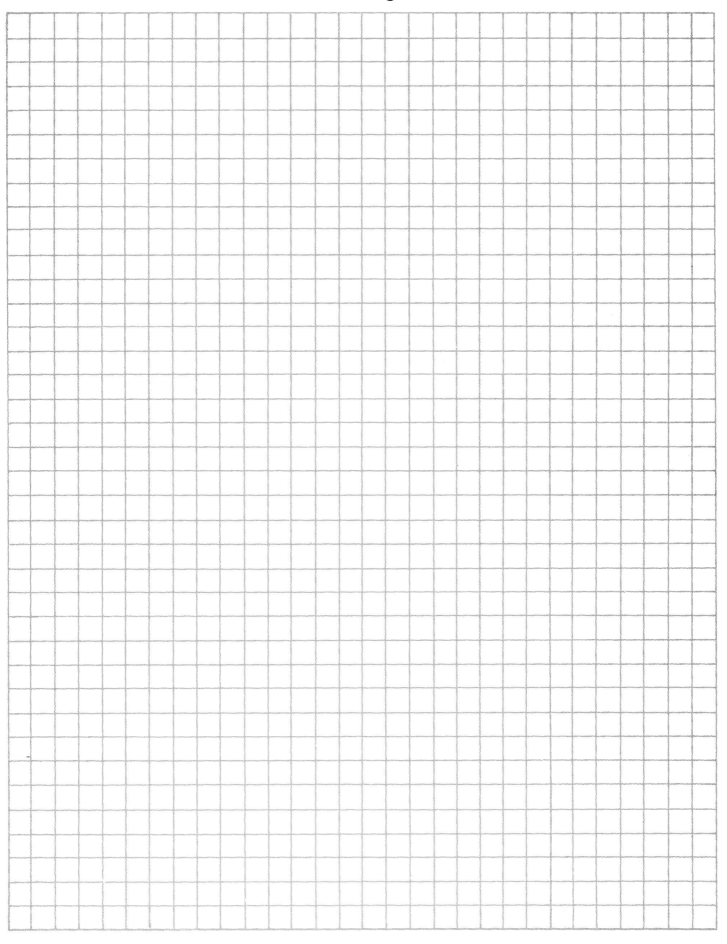

TABLE 5-9 System for Stubbing Teeth†

Depth, percent	Lead angle λ, deg
90	30–34.99
80	35–39.99
70	40–45

†Other systems for stubbing gear teeth such as reducing the depth by 2 percent per degree of lead angle over 30° are also in common use.

listed in Table 5-10. However, other acceptable practices are used by several manufacturers.

Table 5-9 presents a system for stubbing teeth to be used in conjunction with pressure angles in Table 5-8 for lead angles 30° and above.

TOOTH THICKNESS. The gear-tooth normal thickness preferably should be not less than half the normal pitch at the mean of the working depth. In view of the lower-strength material normally used for the gear, it is the practice of some manufacturers to make the gear tooth appreciably thicker than the worm thread. The extent to which this procedure can be followed is limited by the necessity for providing adequate land thickness at the thread peaks.

TOOTH OR THREAD FORMS. The most important detail of the worm thread is that is must be conjugate to that of the gear tooth. The thread form varies with individual manufacturer's practices and may be anything between the extremes of a straight side in the normal section of an involute helicoid.

5-7-8 Gear Blank Dimensions

FACE WIDTH. The effective face of a worm gear varies with the nominal pitch diameter of the worm and the depth of thread. The formula for gear face given in

TABLE 5-10 Dimensions Common to Both Worm and Gear†

Quantity	Symbol	Formula
Addendum	a	$0.3183 p_x$
Whole depth	h_t	$0.6866 p_x$
Working depth	h_k	$0.6366 p_x$
Center distance‡	C	$\dfrac{D + d}{2}$

†Recommended for lead angles less than 30°. See Table 5-9 for others.

‡Nominal, where addenda and dedenda of worm and gear are equal.

Notes · Drawings · Ideas

Table 5-6 is based on the maximum effective face of a worm gear (the length of a tangent to the mean worm diameter) between the points where it is intersected by the outside diameter of the worm. Any additional face is of very little value and is wasteful of material.

DIAMETER INCREMENT. This is the amount that is added to the throat diameter of the gear to obtain the outside diameter. The magnitude of this increment is not critical and may vary with manufacturer's practice. Normal practice is to use approximately one addendum. It is general practice to round the outside diameter to the nearest fraction of an inch.

The sharp corners at the point where gear face and outside diameter intersect should be removed by the use of either a chamfer or radius, as shown in Fig. 5-7. Rim thickness are generally equal to or slightly greater than the whole depth of the teeth.

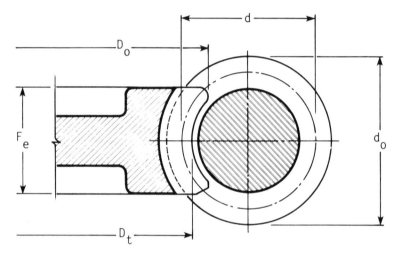

FIG. 5-7 Section of worm and gear. Note that corners of gear teeth are usually rounded, as shown above the gear centerline; they may, however, be chamfered, as shown below.

5-7-9 Worm Face

The face or length of the worm should be such that it extends beyond the point where contact between its threads and the teeth of the gear begin. Unlike with spur and helical gears, the pressure angle of a worm gear varies along the length of the tooth and becomes quite low on the leaving, or recess side. This causes contact to occur on the worm almost to the point where the outside diameter of the worm intersects the throat diameter of the gear.

The formula in Table 5-7 provides a conservative value of the worm face width and is based on intersection of worm outside diameter with gear throat diameter.

More exact worm face widths may be determined by detailed calculations or layouts which take into consideration the face width of the gear and fix more definitely the extent of contact along the worm threads.

Good practice includes the breaking or rounding of the sharp edge of the worm

TABLE 5-11 Range of Recommended Gear-Tooth Numbers

Center distance, in	No. teeth
2	24–40
3	24–50
4	30–50
8	40–60
15	50–60
20	50–70
24	60–80

threads at the end of the worm face. This procedure is particularly important where the worm face is less than provided for in the formula in Table 5-7.

5-7-10 Bored Worm Blanks

Where it is necessary to use a bored worm, the blank is normally designed with a key seat for driving purposes. The thickness of material between the worm root and the key seat should be at least $0.5h_t$. This is a general recommendation which is governed to some extent by whether the blank is hardened or unhardened. An increase in this amount may be necessary if the blank is hardened, particularly if a case-hardening process is used.

5-8 DOUBLE-ENVELOPING GEAR SETS[†]

5-8-1 Number of Teeth in Gear

The number of teeth for the gear is influenced to a large extent by center distance. The recommended number of teeth for various center distances is listed in Table 5-11. Should special considerations indicate a requirement for fewer teeth, it is advisable to consult a manufacturer of this type of gearing before you complete the design. For multiple-thread worms, the number of teeth in the gear should be within the limits listed in Table 5-11. The maximum number of teeth for single-threaded worms is limited only by the machines available for cutting gear sets and manufacturing tooling.

5-8-2 Number of Threads in Worm

The minimum number of teeth in the gear and the ratio determine the number of threads for the worm. Generally, one to nine threads are used. In special cases, a larger number of threads may be required.

[†]See Ref. [5-2].

5-8-3 Gear Ratio

The gear ratio is the quotient of the number of teeth in the gear and the number of threads in the worm. Either prime or even ratios may be used; however, hob life is improved with even ratios.

5-8-4 Pitch

It is recommended that pitch be specified in the axial section. Pitch is the result of design proportions.

5-8-5 Worm Root Diameter

The recommended root diameter for the worm is

$$d_R = \frac{C^{0.875}}{3} \qquad (5\text{-}23)$$

It is desirable that the root diameter be not less than that indicated by this formula, even where the worm threads are cut integral with the shaft. For ratios less than 8/1, the worm root diameter may be increased. This increase may vary from zero for an 8/1 ratio plus 15 percent for a 3/1 ratio.

5-8-6 Worm Pitch Diameter

The pitch diameter of the worm is assumed to be at the mean of the working depth at the center of the worm and is so considered for all calculations. The approximate worm pitch diameter is

$$d = \frac{C^{0.875}}{2.2}$$

and the corresponding root diameter is

$$d_R = d - 2b_G$$

where b_G is the dedendum of gear teeth in inches. Compare this root diameter with that given by Eq. (5-23). If it does not agree, alter the pitch diameter until the root diameter is within the desired limits.

Where horsepower rating is not a factor, there is no limitation regarding pitch diameter of the worm. Where efficiency is not as important as strength or load-carrying capacity, increasing the worm root diameter and gear face width will result in greater capacity.

5-8-7 Base Circle

The base circle may be secured from a layout such as is shown in Fig. 5-10. The normal pressure angle is always 20°. The axial pressure angle may be obtained from

$$\phi_x = \tan^{-1} \frac{\tan \phi_n}{\cos \lambda} \tag{5-24}$$

Once the centerline of the worm and gear, the vertical centerline, and the gear pitch circle are laid out, measure along the common worm and gear pitch circle to the right or left of the vertical centerline an amount equal to one-fourth the axial circular pitch p_x. Through the point thus established and at an angle to the vertical centerline equal to the axial pressure angle ϕ_x, extend a line upward. A circle tangent to this line and concentric to the gear axis is the base circle. Adjust this diameter to the nearest 0.01 in. The formula for figuring the base circle diameter is

$$D_b = D \sin \left(\phi_x + \frac{90°}{N_G} \right) \tag{5-25}$$

5-8-8 Tooth Depth Proportions

Formulas for figuring the whole depth, working depth, and dedendum of gear teeth are found in Table 5-12. Note that the working depth is based on the normal circular pitch and so varies for a given axial pitch.

It is common practice in double-enveloping worm gears to proportion the gear tooth and worm thread thickness as follows: The gear tooth thickness is 55 percent of the circular pitch, and the worm thread thickness is 45 percent of the circular pitch. The backlash in the gear set is subtracted from the worm thread thickness. This practice has been followed to secure greater tooth strength in the gear, which is the weaker member.

TABLE 5-12 Recommended Worm Tooth Dimensions

Quantity	Formula
Length of flat on outside diameter of worm, in	$f = \dfrac{p_x}{5.5}$
Whole depth of tooth	$h_t = \dfrac{p_n}{2}$
Working depth of tooth	$h_k = 0.9 h_t$
Dedendum	$b_G = 0.611 h_k$
Normal pressure angle	$\phi_n = 20°$
Axial pressure angle at center of worm	$\phi_x = \tan^{-1} \dfrac{\tan \phi_n}{\cos \lambda_c}$
Lead angle at center of worm	$\lambda_c = \tan^{-1} \dfrac{D}{m_G d}$

5-8-9 Tooth or Thread Forms

The thread form is usually straight in the axial section, but any other form may be used. Since there is no rolling action up or down the flanks, the form is unimportant, except that it must be the same on the worm and hob. The straight-sided tooth in the axial section provides the greatest ease of manufacture and checking, of both the gear sets and the cutting tools.

5-8-10 Worm Length

The effective length of the worm thread should be the base circle diameter minus $0.10C$ for lead angles up to and including 20° and minus $0.20C$ to $0.30C$ for lead angles from 20 to 45°. The principal reasons for altering this length is to secure the proper amount of worm thread overlap. The overlap should be a distance along the worm thread greater than the face width of the gear. The worm thread extending beyond the effective length must be relieved to prevent interference.

The outside diameter of the worm equals the diameter at the tip of the worm thread at the effective length.

A formula for computing the flat on the outside diameter of the worm at the effective length is given in Table 5-12; the worm face equals the effective length plus twice the flat. The worm face angle is generally 45°.

5-8-11 Gear Blank Dimensions

The face width of the gear should be equal to the root diameter of the worm. Additional face width will not add proportional capacity and is wasteful of material. Where gear sets are to be used at less than their rated horsepower, the face width may be reduced in proportion.

The gear outside diameter may be the point at which the gear face angle intersects the gear throat radius, or any desired amount less, except not less than the throat diameter. The gear throat diameter equals the gear pitch diameter plus one working depth.

There are generally three types of gear blanks in use: those having the hub integral, those flanged and counterbored for a bolted spider, and those having a through bore and fastened by setscrews (or bolts) inserted in drilled and tapped holes located half in the joint between the blank and spider. In all designs, the thickness of metal beneath the teeth should be $1\frac{1}{4}$ to $1\frac{1}{2}$ times the whole depth of the tooth.

5-8-12 Materials[†]

Most of the rating standards are based on the use of worms made from a through-hardened, high-carbon steel heat-treated to 32 to 38 R_C. Where case-hardened worms are employed, somewhat higher ratings may be used.

Many high-strength gear materials (such as aluminum, heat-treated aluminum, and nickel bronzes) are used for slow speeds and heavy loads at higher ratings than shown in [5-3].

---

†See Ref. [5-3].

Notes · Drawings · Ideas

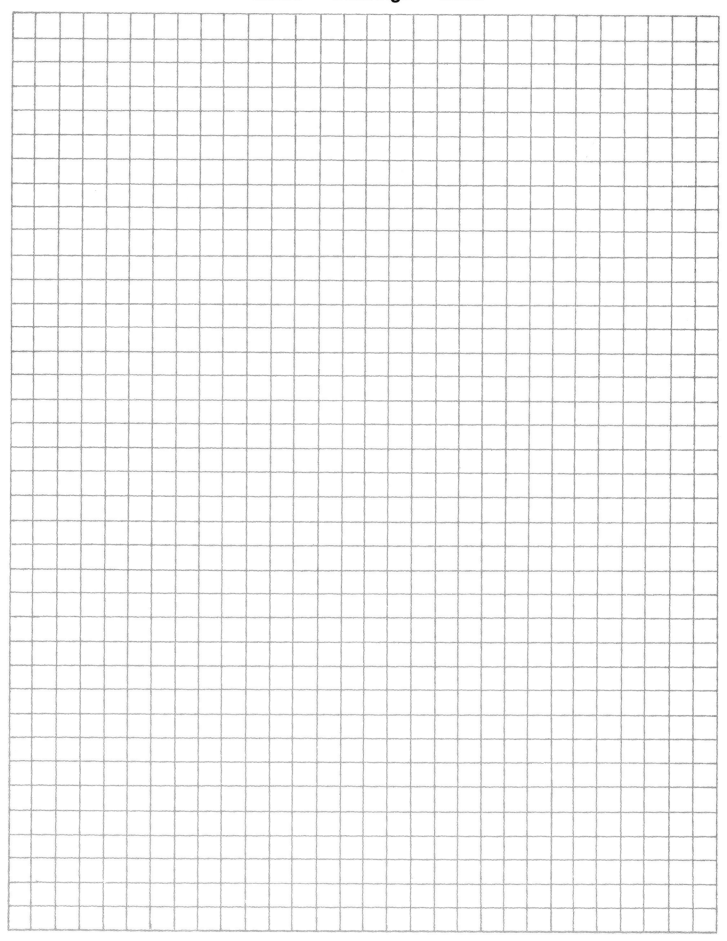

REFERENCES

5-1 "Design of General Industrial Coarse Pitch Cylindrical Worm Gearing," AGMA Standard 341.02-R1970.

5-2 "Design of General Industrial Double-Enveloping Worm Gears," AGMA Standard 342.02-R1976.

5-3 Practices for Single, Double and Triple Reduction Double-Enveloping Worm and Helical-Worm Speed Reducers," AGMA Standard 441.03-1978.

INDEX

Acme thread(s), 3
  standard sizes, 5
  stub, 4
  width of flats, 6
Addendum, 24, 37, 113
American Gear Manufacturer's Association
    (AGMA), 27, 61
  address of, 69
  standards of, 108
Application factor, 28

Back driving, in power ball screws, 19
Backlash, 24, 37
  in power screws, 12, 14
Ball screws, 114
Base circle, 24
Base pitch, 27, 113
Basic load rating, in power screws, 15
Basic static-thrust capacity, in power screws, 15
Bearing stress, in threads, 6
Bending stress, in threads, 6
Bevel gearing:
  allowable stresses, 73, 102, 104
  application requirements, 44
  contact of, 42
  cutter sizes, 54
  estimating sizes of, 46
  face width of, 50
  materials, 54
  pitch, 51
  pressure angle, 53
  radial and thrust forces, 80
  scoring resistance, 74
  selection, 46
  service factors, 71
  testing, 42
  tooth dimensions, 55
  tooth numbers, 48
  tooth taper, 55
Buckling load of threads, 5

Capacity, basic static thrust, power ball
    screws, 15
Case depth, 156
Circumferential stress in a nut, 9
Clearance, 24, 36
Column end conditions for power screws, 8
Cone distance, 37
Contact of gear teeth, 94
Contact lines in gearing, 112
Contact loading, 156

Contact ratio, 27, 112
  profile, 111
Contact stress, 28
Control gear, 39
Cooling of housing, 196
Critical load, in threads, 5
Crown, 39
Cutter sizes, 54

Dedendum, 24, 113
Dedendum angle, 40
Diametral pitch, 24, 40
  normal, 113
Durability, of gearing, 116

Elastic coefficient, 154
  for gears, 154
Elastohydrodynamic (EHD) film, 170
End conditions, column, in power screws, 8
Equivalent load, in power screws, 15
Euler equation for threads, 5

Face angle, 40
Face-contact ratio, 112
Face width, 39, 51, 121
Flash temperature, 176
Force analysis, of bevel gearing, 80
Friction:
  coefficients of, 184
  in power screws, 8
Frosting, 174

Gear noise, 111
Gear ratio, 40
Gearing (*see specific gearing*)
Gears:
  allowable bending stress number, 34
  allowable contact stress number, 30
  bending strength, 32
  bevel and hypoid, 36
  elastic coefficient, 77
  generation methods, 41
  load-distribution factors, 31
  pitting resistance, 28
  power rating, 34
  surface durability, 28
  (*see also specific type of gear*)
Generating rack, 44
Geometry factors for helical gears, 128

Hand of spiral, 77
Hardening, 165

Hardness ratio factor, 164
Heating of housing, 196
Helical gears:
  allowable stresses, 156
  geometry factors, 128
  hardness ratio factors, 166
  life factors, 165
  load capacity, 109
  manufacturing, 111
  noise, 109
  service factors, 118–128
  strength and durability, 116
  temperature factors, 167
  types of, 109
Helix, 109
Herringbone gears, 109
Hoop stress, in a nut, 9
Hypoid offset, 40–52
Hypoid sizes, 60

J. B. Johnson formula for power screws, 5

Lead angle, 181
  in power screws, 10
Life factors, 165
Line of action, 113
Line of centers, 24
Load-distribution factors for gears, 31
Load rating, of power screws, 15
Loads, equivalent, ball power screws, 15
Lubrication:
  of bevel gearing, 92
  of power screws, 10

Materials, for gears, 54
Module, definition, 24

Normal plane, 111
Nuts, circumferential stress in, 9

Overhauling in ball screws, 19
Overload, 71
Overload factors, 71

Parabolic buckling formula, 5
Pinion, 24
Pitch:
  axial, 190
  base, 27
  circle, 24
  circular, 24
  diameter, 24

211

Pitch (*Cont.*)
    diametral, 24, 40, 113
Pitch angle, 40
Pitting, 28
Pitting resistance of gear teeth, 28
Power ball screws, 14
Power rating, 32
Power screws:
    backlash, 13
    efficiency, 10
    lubrication, 13
    rating, 15
Pressure angle, 24, 40, 114, 200
Pressure line, 24

Reliability factors, 159, 166
Reversing screws, 19
Rim thickness, 122
Roller screws, 19
Root angle, 40
Root circle, 24

Scoring, 74, 174
Screw threads, 3

Screws:
    power (*see* Power screws)
    reversing, 19
    roller, 19
Scuffing, 174
Self-locking, 181
Service factors, for bevel gears, 71
Shaft angle, 38, 40
Shear stress, in screw threads, 9
Square thread, 3
Static-thrust capacity of power screws, 15
Straddle mounting, 72
Strength, of gearing, 116
Stress(es):
    allowable, 156
    bearing, 6
    bending, 6
    contact (*see* Contact stress)
    hoop, in nut, 9
    permissible, 102
Stress number, 28, 116
Stub Acme thread, 4
Surface durability, 28, 46
    of gears, 28
Surface texture, 173

Tangential force, 41
Temperature considerations, 167
Temperature rise, of housing, 196
Threads, 3
    buckling of, 5
Thrust forces in bevel gearing, 80
Tooth depth, 200
Tooth systems, 26
Tooth thickness, 113
Transmitted load, 27
Transverse plane, 111

Velocity, pitch-line, 28

Whole depth, 24
Worm, 181
Worm gearing:
    pitch diameters, 200
    speed ratio, 198
    tooth numbers, 198
    tooth proportions, 200

Zerol gears, 37